The WATERSHED PROJECT MANAGEMENT Guide

The WATERSHED PROJECT MANAGEMENT Guide

Thomas E. Davenport

LEWIS PUBLISHERS

A CRC Press Company
Boca Raton   London   New York   Washington, D.C.

Library of Congress Cataloging-in-Publication Data

Davenport, Thomas E.
   The watershed project management guide / Thomas E. Davenport.
      p. cm.
   Includes bibliographical references (p.).
   ISBN 1-58716-092-7 (alk. paper)
   1. Watershed management. I. Title.

TC409 .D297 2002
333.73—dc21   2002067143

         Catalog record is available from the Library of Congress

This book contains information obtained from authentic and highly regarded sources. Reprinted material is quoted with permission, and sources are indicated. A wide variety of references are listed. Reasonable efforts have been made to publish reliable data and information, but the author and the publisher cannot assume responsibility for the validity of all materials or for the consequences of their use.

Neither this book nor any part may be reproduced or transmitted in any form or by any means, electronic or mechanical, including photocopying, microfilming, and recording, or by any information storage or retrieval system, without prior permission in writing from the publisher.

The consent of CRC Press LLC does not extend to copying for general distribution, for promotion, for creating new works, or for resale. Specific permission must be obtained in writing from CRC Press LLC for such copying.

Direct all inquiries to CRC Press LLC, 2000 N.W. Corporate Blvd., Boca Raton, Florida 33431.

Trademark Notice: Product or corporate names may be trademarks or registered trademarks, and are used only for identification and explanation, without intent to infringe.

Visit the CRC Press Web site at www.crcpress.com

© 2003 by CRC Press LLC
Lewis Publishers is an imprint of CRC Press LLC

No claim to original U.S. Government works
International Standard Book Number 1-58716-092-7
Library of Congress Card Number 2002067143
Printed in the United States of America 1 2 3 4 5 6 7 8 9 0
Printed on acid-free paper

Dedication

This book is dedicated to my children, Ethan and Emily, who have had a few fatherless occasions because I was confused over career and family, and to Mom and Dad, whose common sense and support have helped me immensely.

Preface

We all live in watersheds — the area of land that drains into a wetland, stream, or lake. Everything we do affects the quality of that wetland, stream, or lake downstream. We can use most of our lakes and streams and enjoy our wetlands; however, they are becoming more stressed and less desirable because of increased and competing uses and continued accelerated degradation from land and air pollution. The public wants to stop this downward trend.

Many people are becoming locally involved in restoring and protecting the water resources they value. In response to the public's activities, all levels of government are becoming more involved in supporting local efforts to protect or restore water resources through watershed management. Scientific and technical expertise provided by nongovernmental organizations, government agencies, and consultants play a vital role in helping local watershed management efforts.

Local communities, tribes, and states play an increasingly crucial role in designing and implementing federal agencies' water-quality agenda. The public policy debate for federal involvement both at this point and in the future is not whether watershed-based efforts are here to stay, but what form they will take: problem solving, stewardship, or a combination of both. Hopefully they will be based on the locally identified environmental issues and needs, rather than politics or budget constraints. A style of management that emphasizes people getting together cooperatively to solve shared problems seems like a common-sense approach and should be supported by public policy.

Keep in mind that watershed management is largely a process of working with people to solve their problems. Watershed management can be difficult to regulate; regulations can prevent people from doing harmful things, but they provide no incentive for people to do beneficial things. The public must want to become involved for watershed management to succeed. The goal of the watershed management process is "to ensure that water and related resources are managed to provide for the environmental, social, and economic well-being of the stakeholders." Locally led watershed management focuses government agencies' efforts on working with partners solving problems, not running programs. Clearly, watershed management requires the "Ps": planning, partnership, and participation by stakeholders.

The target audience for *The Watershed Project Management Guide* is what I call "watershed practitioners and those who support them." This is not a theoretical or inspirational book, but a practical one. It promotes the wise use of available resources to aid in the environmental improvement of the watersheds in which we live. A major barrier to successful restoration and management of our nation's water resources is a lack of knowledge about watershed processes. This guide is intended to give science and social information that can aid efforts to protect and restore community watersheds. For example, watershed management is most effective where there is active citizen involvement. Active stakeholder involvement ensures that ideas of the people most affected by the outcome are included at the front end of the watershed management process.

Because there are so many different watershed management processes, this book is only a guide to the general or universal principles applicable to all watershed management efforts. Individually, some distinguishing characteristics are not new, and in fact they have been part of land and water management for decades, but in combination they describe a new approach for addressing complex water and related management issues. The resulting four-step process (assessment, planning, implementation, and evaluation) relies on partnership leadership, adequate monitoring, and outreach. This book outlines the four-step process for successful watershed management, summarizes available information and approaches, and provides references for their use. I have tried to emphasize the practical problems (lack of information/data, insufficient time, competing demands) that the practitioner must face in attempting to develop and implement a watershed management approach. Practical examples are used to illustrate the benefits and problems of watershed management and the process. These examples are to help practitioners learn what works and what does not, based on case studies or personal experience.

The entire four-step process is important because a major barrier to watershed management is the perceived endpoint, namely, the watershed plan versus natural resource restoration or protection. For many watershed efforts the completion and approval of the watershed plan ended the effort. In reality the watershed plan is just one step in a dynamic process to restore and protect water quality. Another issue is the focus of the watershed management effort. Historically, the main focus of watershed management activities has been on restoration of impaired waterbodies. Not losing sight is important: in fact, a small investment in pollution prevention today will prevent far more costly pollution-related damage in the future. The process outlined in this book is applicable for both restoration and prevention projects. This book focuses on the complexities of the watershed management process, the watershed partnership's role in the processes, and what needs to be done next.

In this book I seek to impart not merely information and knowledge of what I think watershed management is, but an understanding of the underlying philosophy. I view watershed management as a rational mechanism to focus local behavior. The words "rational," "mechanism," and "local" are the key ideas of this paradigm. In this view, the decision to initiate a watershed management effort ought to be rational, in the sense that it is based on estimated community benefits with the proper management of the water resource and costs associated with these activities. Next, the effort ought to be functional in the sense that it is initiated to achieve some specific water resource goal. Nevertheless, this does not mean other complementary natural resource or social goals cannot be developed and included in the overall watershed management effort since one goal is the consensus overall priority. Finally, watershed management should be local, in the sense that its objectives should be to advance the interests of stakeholders and to engage the entire watershed community in the effort.

For this process to be successful, government needs to play a supporting role for the local efforts. It is important to note the two most important characteristics for government organizations involved in working with citizen watershed management partnerships to exhibit are integrity and sustainability. Due to the time-consuming nature of watershed management, government agencies need to make a substantial long-term commitment when getting involved with citizen groups. Making this type of commitment is difficult for most government agencies due to politics, changing priorities, staff turnover, and lack of understanding by the decision maker.

To fully understand my philosophy you are required to view the logical structure, that is, the way concepts and ideas relate to each other, and how they are derived from other concepts or ideas. Then you compare this view with the existing conditions and accept or reject the philosophy.

I have kept technical jargon to a minimum to help the reader grasp important points without stumbling over the words. Where appropriate, the reader is directed to specific resources and references for further information. English units are used except for a few terms that are almost always reported in metric.

The material presented here fulfills the following purposes:

1. To describe the four-step process for developing a watershed specific management plan
2. To help identify, describe, and define water resource problems and issues
3. To provide tools, approaches, and information that can be used in watershed management plan development and implementation
4. To highlight how to implement a watershed management plan and evaluate its effectiveness

> *The secret of getting ahead is getting started. The secret of getting started is breaking your compelling overwhelming tasks into small manageable tasks, and then starting. Onto the first one.*
>
> — **Mark Twain**

Acknowledgments

This book, as modest as it is, is the result of more than two decades of experience in the field. I have learned from a lot of individuals, my successes and failures. The combination of ideas, insights, and experience has molded and contributed to what this book covers. I am very grateful to all of them.

The older generation includes James Meek, James Pagenkopf, Old Walt Poole, Chuck Sutfin, Sue Elston, and Drs. Nowak and Brady — whose mistakes and guidance I found enlightening. My colleagues include the Kirschner Brothers, Paul Thomas, Tom Schueler, Tim Icke, Sarah Lehmann, Bob Kirschner, George Townsend, Steven Dressing, and Ernie Lopez — whose daily support has been needed and appreciated. Special thanks go to The Terrene Institute and Conservation Technology Information Center for most of the figures, Melissa DeSantis and Charlie MacPherson for the graphics help, Diana Allen for creative ideas, Hye Yeoung Kwon for the Ben Franklin quote, and Karen Scanlon for her work and ideas on watershed training. Special thanks also go to Emily Ann Davenport for her production assistance.

I would also like to acknowledge Ethan Davenport, who, when I tried to explain what I did for a living, looked confused and said he would wait for the book.

It is important to thank all the individuals at CRC Press, especially Michele Berman and Marsha Hecht, project editors, and Pat Roberson, production manager, who made this possible.

If I have missed anyone, I am sorry.

Thomas E. Davenport
Certified Professional Erosion and
Sediment Control Specialist

Abstract

The key question for individuals involved in managing watersheds is, "What is an effective process to integrate science, policy, and public participation to manage water resources effectively?" There is a need for a flexible process to take advantage of emerging opportunities and the unique circumstances of each watershed. This book presents a four-phase approach to watershed management that is based on a collaborative process that responds to common needs and goals. It utilizes assessments and decision processes that are based on a combination of biophysical, social, and economic information, plus local knowledge. The "recommended process" consists of a series of four basic phases; assessment, planning, implementation and evaluation built on stakeholder involvement, social capacity building, and adequate monitoring. The four-phase approach assists practitioners in developing a plan that can be used to implement a management strategy to meet the load allocations required by an approved TMDL, the goals of a source water protection plan, USDA conservation programs or Section 319 project.

Author

Thomas E. Davenport, CPESC #174, is an environmental scientist for the U.S. Environmental Protection Agency (USEPA) and was designated as the National Nonpoint Source Expert in 1994. For the 10 years prior to this designation, he served as the USEPA's Region 5 coordinator for the Nonpoint Source and Clean Lakes Programs. Present duties include the Water Program Lead for the Great Lakes/Baltic Seas and Three Rivers, Three Countries Watershed Capacity Building Projects.

Davenport received a B.S. in forestry and natural resource management from the University of Wisconsin, Stevens Point, in 1977 and an M.S. from the University of Washington in forest hydrology in 1981. In 1982, he received an M.P.A. from Sangamon State University.

Davenport has received seven bronze medals from the USEPA for outstanding contributions for various activities related to nonpoint source, lake restoration, and watershed management. In June 1984 he received a Certificate of Merit from the U.S. Department of Agriculture for assistance related to the development and implementation of the National Farm*A*System. The author received the President's Award from the Minnesota Lakes Association.

Davenport has participated in numerous international and national meetings, has presented five invitational watershed management workshops at international meetings, and has published over 40 papers, book chapters, and project reports. He is currently on the editorial board of the Center for Watershed Protection's technical journal, *Watershed Protection Techniques*, and he is an associate editor of research for the *Journal of Soil and Water Conservation* for the Soil and Water Conservation Society. Previously, Davenport was on the editorial board for the *Journal of Soil and Water Conservation*.

Acronyms

AgNPS Agricultural Nonpoint Source Pollution Model
ARS Agricultural Research Service
BASINS better assessment science integrating point and nonpoint sources
BAT best available technology
BLM Bureau of Land Management
BMP best management practice
BOD biochemical oxygen demand
CAC citizen advisory committee
CAFO concentrated animal feeding operation
CERCLA Comprehensive Environmental Response, Compensation, and Liability Act
cfs cubic feet per second
COD chemical oxygen demand
CMC Center for Marine Conservation
CRP Conservation Reserve Program
CTIC Conservation Technology Information Center
CWA Clean Water Act
CWAP Clean Water Action Plan
CZARA Coastal Zone Act Reauthorization Amendment
CZMA Coastal Zone Management Act
DOE Department of Energy
DOI Department of the Interior
EIS environmental impact statement
EMC event mean concentration
EPA Environmental Protection Agency
EQIP Environmental Quality Incentive Program
ESA Endangered Species Act
ET evapotranspiration
FWS Fish and Wildlife Service
GIS geographic information system
GPD gallons per day
GRASS geographic resources analysis support system
GWLF general watershed loading functions
IBI Index of Biological Integrity
IJC International Joint Commission
IPM integrated pest management
ISTS individual sewage treatment system
ISWS Illinois State Water Survey
LA load allocation
lb/ac pounds per acre
KASA knowledge, ability, skills, and aspirations

mgd million gallons per day
MOA memorandum of agreement
MOS margin of safety
MOU memorandum of understanding
N nitrogen
NAWQA National Water Quality Assessment Program
NOAA National Oceanic and Atmospheric Administration
NO_2+NO_3 nitrite plus nitrate (dissolved in water)
NO_3 nitrate
NPDES National Pollutant Discharge Elimination System
NPS nonpoint source
NRC National Research Council
NRCS Natural Resources Conservation Service (formerly SCS)
NWI National Wetlands Inventory
O&M operation and maintenance
OLRPC Otter Lake Resource Planning Committee
OST on-site sewage treatment
PCB polychlorinated biphenyls
PCS permit compliance system
POTW publicly owned treatment works
PUD planned unit development
PWD Public Works Department
Q discharge
RCWP Rural Clean Water Program
RMS resource management system
SCS Soil Conservation Service
SRF state revolving fund
SSOs sanitary sewer overflows
STORET Storage and Retrieval of U.S. Waterways Parametric Data
TAC technical advisory committee
TKN total Kjeldahl nitrogen
TMDL total maximum daily load
TN total nitrogen
TP total phosphorus
TRI Toxics Release Inventory
TSI Trophic State Index
TSS total suspended solids
USACE U.S. Army Corps of Engineers
USDA U.S. Department of Agriculture
USEPA U.S. Environmental Protection Agency
USFS U.S. Forest Service
USGS U.S. Geological Service
USLE universal soil loss equation
WLA waste load allocation
WQS water-quality standards
WRAS watershed restoration action strategy

Table of Contents

Chapter 1 Why Watersheds?..1
1.1 Introduction...1
1.2 Watershed Management...4
      1.2.1 Model Watershed Process..6
1.3 Summary...8
References..8

Chapter 2 Watershed Management Process..9
2.1 Introduction...9
2.2 Watershed Management Process..11
2.3 The Four-Step Process..13
      2.3.1 Assessment and Problem Identification Phase:
            What Is Happening to the Resource?..13
      2.3.2 Planning Phase: What Do We Need to Do?
            How Will We Get There?..15
      2.3.3. Implementation Phase: Making a Difference....................................17
      2.3.4 Evaluation Phase: Did We Make It?...17
2.4 Summary...19
References..19

Chapter 3 Watershed Processes...21
3.1 Introduction...21
3.2 Watershed Units..23
3.3 Drainage Network...24
3.4 Water Cycle..25
      3.4.1 Precipitation..26
      3.4.2 Runoff...28
      3.4.3 Stream Flow...29
      3.4.4 Groundwater...29
3.5 Geomorphologic Processes..30
3.6 Aspects for Management..31
3.7 Land Use...32
      3.7.1 Urban..33
      3.7.2 Agriculture...33
      3.7.3 Atmospheric Deposition..34
References..35

Chapter 4 Partnership Development and Operation 37

4.1 Introduction 37
4.2 Building Partnerships 38
4.3 Organization 41
    4.3.1 Structure 42
    4.3.2 Partnership Operations 49
4.4 Institutional Aspects 56
4.5 Political Involvement 57
4.6 Conclusion 59
References 60

Chapter 5 Assessment and Problem Identification Phase 61

5.1 Introduction 61
5.2 Assessment 63
    5.2.1 Defining the Watershed 65
    5.2.2 Project Scope 67
    5.2.3 Inventory Stage 68
5.3 Analysis 71
    5.3.1 Three-Tier Analysis 74
5.4 Pollutant Source Assessment 77
    5.4.1 Urban 80
    5.4.2 Agriculture 83
5.5 Critical Areas 84
5.6 Pulling It Together 86
References 89

Chapter 6 Plan Development 91

6.1 Introduction 91
6.2 Background 92
6.3 Planning 94
    6.3.1 Planning Phase 96
    6.3.2 Priority Setting 98
    6.3.3 Watershed Strategy Development 100
6.4 Implementation Approaches 102
    6.4.1 Agriculture 103
    6.4.2 Urban 107
    6.4.3 Urban Riparian and Wetland Management 113
6.5 Restoration Opportunity 115
6.6 Conclusion 117
References 120

Chapter 7 The Watershed Management Plan 123

7.1 Introduction 123

7.2 Plan Elements 123
7.3 Plan Format 126
7.4 Funding 126
References 128

Chapter 8 Implementation 129

8.1 Introduction 129
8.2 Institutional Aspects of Implementation 130
8.3 Implementation Scale 132
    8.3.1 Community Scale 132
    8.3.2 Field Scale 135
8.4 Management Practice Aspects 138
8.5 Volunteer Contributions 139
8.6 Implementation Funding 139
8.7 Implementation Dynamics 139
References 140

Chapter 9 Evaluation 141

9.1 Introduction 141
9.2 Evaluation Types 144
9.3 Evaluation Barriers 149
9.4 Evaluation Levels 150
9.5 Alternative Evaluation Approach 156
9.6 Evaluation Plan 158
9.7 Evaluation Reporting 159
9.8 Summary 160
References 161

Chapter 10 Monitoring 163

10.1 Introduction 163
10.2 Watershed Monitoring 164
10.3 Types of Monitoring 167
10.4 Monitoring Purposes 168
10.5 Scale of Monitoring 175
    10.5.1 Field 175
    10.5.2 Subwatershed 175
    10.5.3 Watershed 176
10.6 Volunteer Monitoring 176
10.7 Development of a Monitoring Effort 177
10.8 Reporting 179
References 180

Chapter 11 Models 183

11.1 Introduction 183
11.2 Model Category 184
11.3 Model Application 186
11.4 Model Selection 190
11.5 Model Documentation Plan 193
11.6 Geographic Information Systems 193
References 200

Chapter 12 Social Capacity Building 203

12.1 Introduction 203
      12.1.1 Public and Stakeholders 205
12.2 Outreach 205
12.3 Education 209
      12.3.1 Knowledge Stage 211
      12.3.2 Persuasion Stage 212
      12.3.3 Implementation Stage 213
      12.3.4 Evaluation (Confirmation) Stage 213
12.4 Communications Plan 213
12.5 Public Participation 218
12.6 Planning for Successful Participation 223
12.7 Annual Review 224
References 224

Chapter 13 Conclusions 227

Glossary 231

Index 251

1 Why Watersheds?

One generation plants the trees; another gets the shade.

— **Chinese Proverb**

1.1 INTRODUCTION

Like planting the shade tree, watershed management must start today so future generations can reap the benefits of clean water. America's water resources are critical, providing drinking water, food, and recreational opportunities to all. These benefits must be preserved for society's survival, not just for those who can afford to live in relatively undisturbed or pristine areas. The U.S. is tremendously wealthy in aquatic resources with more than 3.6 million miles of streams (rivers), 41.6 million acres of lakes (reservoirs and ponds), 90,400 miles^2 of estuaries, approximately 274 million acres of wetlands, 72,000 miles of coastal shorelines, and abundant groundwater supplies. Overall these resources provide great economic, social, and cultural value. The availability of water resources and other raw materials provided the foundation for the economic growth of the United States, resulting in America becoming the world's largest producer and consumer. However, this production of goods and services has had detrimental environmental impacts. For example, removing forest cover and installing a drainage system to allow for farming create downstream flooding due to altering the natural processes of upstream water storage.

Pollution and environmental degradation are products of living. According to *USA Today*,[1] "More people die each year from unsafe water than from all forms of violence, including war. More than a billion people — one in every five on earth — do not have access to safe drinking water." Society has adopted a goal of minimizing pollution and restoring those areas degraded by pollution. The economic and physical conditions relative to a watershed's functions have little correlation with patterns of water consumption.[2] Historically, we have attempted to reduce pollution and address environmental problems with single-purpose, piecemeal approaches. While these single-purpose efforts can be effective in addressing that particular issue, the "solutions" have the potential to create unintentional problems elsewhere. For example, to decrease downtown flooding, drainage networks have become so regulated they can no longer support other resource objectives such as fishery management. Agencies set regulatory standards for variables in the environment, based on the minimum standard necessary to meet designated bureaucratic, technical, or environmental objectives — or a combination of them.

The use of single-purpose regulations has had the unintended impact of lowering overall environmental quality in some locations. This type of regulatory approach leads to control and treatment strategies that use the minimum standards as targets. In complying with these minimum standards, most efforts interpret the targets as the goal they must achieve rather than viewing them as a minimum from which to build. This regulatory approach by minimum standard has fostered the creation of a minimum environment. If one looks at the implementation of traditional pollution control programs for air, water, and land within the context of political boundaries, one will see the obvious need for an integrated approach to go beyond the minimum. Scientists have found solving one problem may inadvertently create other problems. Implementing conservation tillage to address soil erosion problems associated with surface runoff promotes increased infiltration and the transport of water-soluble chemicals to groundwater — and this results in improved surface-water quality and the potential to degrade groundwater quality.

Nationally much progress has been made in cleaning up impaired waters, but some waters in this country still cannot support the goals of the Clean Water Act. The 1998 National Water Quality Inventory[3] reports that about 40% of the assessed waters do not meet water-quality goals. Impaired waters are not safe for one or more uses, including fishing, swimming, drinking, or supporting aquatic life. The Coastal Alliance[4] reports algae blooms attributed to nonpoint source pollution have caused massive fish kills and impaired human health. Bacterial and chemical pollution threatens drinking water systems, requiring expensive upgrades in treatment capability. Beach closings, contaminated fish and shellfish, property damage, and polluted drinking-water supplies from contaminated runoff are costly to communities as well. Entire ecosystems upon which the economy and health of whole regions depend have been decimated. The Everglades are an example of this level of impact. About half the nation's more than 2000 major waterbodies have serious or moderate water-quality problems. Runoff pollution also affects our recreational opportunities by adversely impacting fish and wildlife resources (see Tables 1.1 and 1.2).

According to the EPA,[3] more than 80% of Americans live within 10 miles of a polluted river, lake, or coastal water. Among all pollutants afflicting streams and rivers

TABLE 1.1
Leading Sources* Causing Impairments in Assessed Rivers, Lakes, and Estuaries

Rivers and Streams	Lakes, Ponds, and Reservoirs	Estuaries
Agriculture	Agriculture	Municipal point source
Hydromodification	Hydromodification	Urban runoff/storm sewers
Urban runoff/storm sewers	Urban runoff/storm sewers	Atmospheric deposition

* Excludes unknown, natural, and other source categories.

Source: USEPA, Water Quality Conditions in the United States: A Profile from the 1998 National Water Quality Inventory Report to Congress, EPA-841-F-00–006, USEPA, Washington, D.C., 2000.

TABLE 1.2
Leading Pollutants Causing Impairments in Assessed Rivers, Lakes, and Estuaries

Rivers and Streams	Lakes, Ponds, and Reservoirs	Estuaries
Siltation	Nutrients	Pathogens
Pathogens	Metals	Organic enrichment
Nutrients	Siltation	Metals

Source: USEPA, Water Quality Conditions in the United States: A Profile from the 1998 National Water Quality Inventory Report to Congress, EPA-841-F-00–006, USEPA, Washington, D.C., 2000.

in the U.S., sediment is by far the greatest in terms of volume. Several anthropogenic sources of sediment are likely to enter and degrade our nation's water resources. Agriculture in its varied forms is by far the largest generator of sediment. Agriculture is the most widespread nonpoint source of water pollution across the nation.[3] The most severe agricultural-related problem documented was soil erosion resulting in siltation. Approximately half the total sediment delivered to U.S. lakes and streams is from cropland. The runoff from agricultural lands, carrying large amounts of sediment, scours the stream channel, alters the character of the stream, and affects aquatic life, impairing functions such as photosynthesis, respiration, growth, and reproduction. While the other land uses do not impact water quality to the same degree as agriculture, these impacts can be locally significant and have a greater impact on the overall population. This is especially true for pollutants other than sediment. For example, elevated levels of pathogens are the leading cause for beach closures in the U.S., and the two most common sources are municipal point sources and urban runoff.

The cumulative impacts of human-induced pollution results in significant changes, not only to the immediate drainage system but also to the watershed off which it drains. These cumulative changes include degradation of water quality, decreased water storage capacity, loss of habitat for fish and wildlife, decreased water conveyance capacity, and decreased recreational opportunities. Nonpoint source pollution's diffuse nature causes many problems that do not lend themselves to traditional pollution control approaches. Besides the cumulative impacts of pollution, there is no universally accepted process that integrates competing water uses/issues such as water quality, water supply, flood control, navigation, hydroelectric power generation, fisheries, and recreation with economic and social constraints to maximize benefits and minimize costs and negative impacts to society. The unmet need of economic integration and the cumulative impacts of pollution have been driving forces in initiating watershed management. The usefulness of watersheds is based on the understanding that the quantity and quality at a point on a stream reflect the aggregate of the characteristics of the topographic up-gradient from that point.

A *watershed* is the drainage area for the entire waterbody system, including lakes, streams, wetlands, groundwater, and the land. A watershed contributes the water required to maintain, and most of the pollutants that enter, a waterbody.

1.2 WATERSHED MANAGEMENT

Comprehensive watershed management serves to integrate human influences and natural factors in practical, tangible management units. This type of management is not a new program but is a process to align, coordinate, and build upon existing formal and informal programs. It complements other types of natural resource management such as agricultural and fish and wildlife resource targeting programs. It focuses on hydrologically defined management units — watersheds — rather than on areas defined by political or ecoregion boundaries.

The purpose of *The Watershed Project Management Guide* is to provide a practical and applied guide to watershed management. The inspiration and concepts supporting watershed management projects and the institutional barriers against them are inherited from history and legislation. It is important to realize that the field of watershed management is not new in the U.S. In his will Ben Franklin promoted watershed management:

> After having considered that covering the ground of the city with building and pavements which carry off most of the rain, and prevents its soaking into the earth and renewing and purifying the springs, whence the water of wells must gradually grow worse and be unfit for use, as I find has happened in old cities ... I recommend at the end of the first hundred years, if not done before, the ... city employ a hundred thousand pounds in bringing by pipes water so as to supply the inhabitants.

Even though Franklin recognized the need for watershed management in 1790, the federal government's watershed management roots go back to the Organic Administrative Act of 1897 that created the National Forest System. The dual purposes of this system were preserving favorable conditions of water flows and ensuring continuous timber supply. Since then numerous pieces of legislation have been enacted to cause the evolution of American institutions and programs for water quality such as the Rivers and Harbors Act of 1899 and the Small Watershed Program and Federal Water Pollution Act of 1972.

Early attempts at watershed management were often not fully successful because they had:

- Too narrow a focus (e.g., Small Watershed Program, Organic Administrative Act)
- Too few "hard" tools like water-quality standards and regulations (e.g., Clean Water Act's Section 208 planning)
- A lack of today's information and technology (e.g., remote sensing, global positioning systems, geographic information systems)
- A lack of understanding interrelationships between ecological and human factors

Between the late 1980s and early 1990s a gradual change in watershed management efforts started because of:

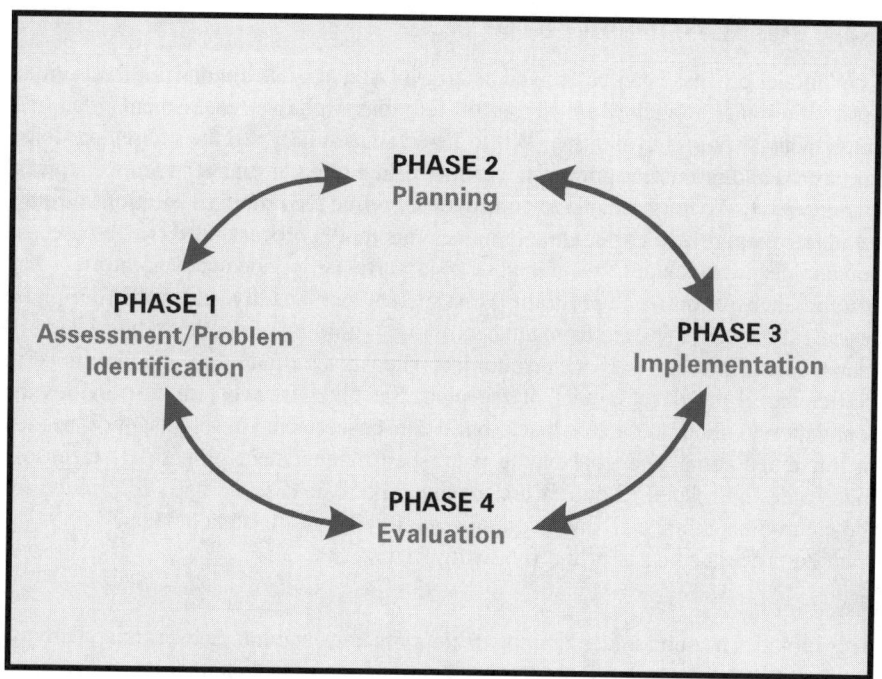

FIGURE 1.1 The four phases of a watershed management model.

A general recognition that many remaining problems were not amenable to traditional environmental regulation (nonpoint sources, habitats, in-place sediments, cumulative impact of minor sources)

A growing sense of need for partnering and the capacity to do so

A better understanding of the interconnectedness of everything

What has evolved since the 1990s is a watershed management process that provides a dynamic and flexible approach to meet changing goals and needs. The four phases of the approach (Figure 1.1) provide a framework to organize information and decision points. The process provides a zooming capability to meet data and analysis needs for each phase. This model process responds to a conscious decision at the local level to attempt to deal with some tough and complex issues:

Using boundaries rather than traditional government jurisdiction
Addressing all pollutant sources
Working with an entire community
Ownership of the decision-making process

The watershed management process is the primary approach for relating science, policy, and public participation to water resources management. Overall, watershed management is most effective if it is simple and logical. The watershed management process and written plans need to be viewed as supporting, not prescriptive, tools.

1.2.1 MODEL WATERSHED PROCESS

The "model process" can be viewed as an evolving approach with many variations. At a minimum, it includes a series of four basic phases: assessment, planning, implementation, and evaluation. While listed sequentially, all the phases are interconnected and are most appropriately viewed in a circular manner such as depicted in Figure 1.1. We must be able to continuously refine the proposed solutions through an iterative approach. Experience indicates this model process needs to be based on stakeholder involvement, management by a partnership, adequate monitoring, and a comprehensive outreach program. Most stakeholders readily accept the four-phase process if they are involved from the beginning. A guiding principle of this watershed management process is that stakeholders need a legitimate early opportunity to participate in the development of the plan. Stakeholder involvement provides the foundation to obtain pubic feedback, build consensuses, and develop support needed in implementation. The goal of the watershed management process is "to ensure that water and related resources are managed on a sustainable basis to provide for the environmental, social, and economic well-being of the stakeholders."

The process consists of the following four stages.

Assessment
This provides a sound understanding of the processes and interactions in and around the watershed. This phase includes the identification of stakeholder concerns, institutional concerns, and the social/economic characteristics of the watershed.

Planning
Planning means the development of an implementation plan based on a diagnostic analysis from stakeholder input and the information and inventories assessments. This phase includes analysis and selection of alternatives and documentation of decisions. In this context, "plan" is a verb, not a noun. The outcome of planning is not intended to be a document; it must include action and direction. The implementation strategy should be the best approach to solving the identified problems and resolving outstanding data needs. The plan is more "bottom-up" than "top-down" and is built on a solid foundation of water-quality standards, regulation, voluntary incentives, education and information, and enforcement with everyone at the table. The watershed management plan does not have to be completed before implementation. Known problems with known solutions should be addressed immediately in this process, to start building success for the overall effort.

Implementation
This is the culmination of the initial assessment and planning efforts, organizing, and information/education activities — bringing together previous efforts and actually doing work. It can include pollution prevention and reduction activities and restoration efforts. Implementation builds off point source controls that have historically been targeted toward compliance with chemical standards representing maximum acceptable pollutant levels or levels that can be exceeded only infrequently and incorporates the management of nonpoint sources of pollution. It is usually funded through a variety of public and private programs and in-kind contributions.

Why Watersheds?

Evaluation

Evaluation starts during the assessment phase and continues after the implementation phase is completed. This phase tells a partnership whether or not it did what it set out to do and if the plan had the impact expected. Evaluations during the implementation phase provide the basis for midcourse corrections based on what is working and what is not. It also provides the basis for celebrating the accomplishments and reporting to the public.

The numerous benefits of this model process can be broken down into three general categories. First, this approach provides a context for integration by using practical, tangible management units. The size of the watershed addressed is tied to problems and local capabilities and interests. It promotes finding common ground and meeting multiple stakeholder needs. The process facilitates the focusing and coordinating of federal and state efforts to address local problems. Second, it promotes a better understanding and appreciation of nature. During the assessment phase, stakeholders gain an understanding of nature's interrelated processes, appreciating how nature's processes can benefit humans and identifying ways they can work with watershed processes. Stakeholders also learn to link human activities to changes in water quality and quantity and to answer the question, "What are they trying to protect or restore?" Last, this model process promotes better management. Using this approach, partnerships forge strong working relationships and use the management plan to promote consistent, continuous management to meet their goals. The management plans developed by partnerships with stakeholder involvement are tailored to local needs, resources, abilities, and will. No two plans should be exactly alike.

The model is presented in a linear format; however, a true watershed management process is circular. New information, new stakeholders, or an ongoing evaluation that shows action taken did not have the expected results can require significant alteration to the existing approach. The process supports a flexible, dynamic approach that can be frequently updated to keep the focus on the goals and progress toward them.

Before partnerships can start on a watershed management process, they have to know if the timing is right for the watershed in question. The Friends of the Chicago River[5] have developed a list of questions to consider when prioritizing watersheds for attention. A modification of Cohn's list is provided here, which is more applicable to a wider range of watersheds and circumstances:

> What is public interest in or knowledge about the watershed? What are the public concerns about water quality, flooding, or other watershed-related issues?
> Is there an existing organization to work with? Are there agencies and organizations that can help promote watershed management?
> What has already been done in the watershed? Have there been previous studies, planning efforts, special-purpose investigations, or demonstration projects?
> Does the watershed have a mix of interested groups and individuals who will become involved in the watershed management process?
> If the watershed is transboundary, are all levels and units of government interested?

Answering these questions will enable a partnership to develop a strategy for moving ahead and developing a watershed management plan.

1.3 SUMMARY

Because water integrates all landscapes — urban and nonurban — the land use and its integrity are essential to the welfare of humans and the environment. While it would be good for society if aquatic ecologists and water resource planners were in the forefront of watershed management, instead the locals are trying to lead the way — and the professionals must catch up.

REFERENCES

1. Water Unsafe in Much of World, *USA Today*, May 17, 2001, Section A, p. 1.
2. Ciriacy-Wantrup, S.V., Philosophy and objectives of watershed policy, in *Economics of Watershed Planning*, Iowa State University Press, Ames, 1961.
3. USEPA, Water Quality Conditions in the United States: A Profile from the 1998 National Water Quality Inventory Report to Congress, EPA-841-F-00–006, USEPA, Washington, D.C., 2000.
4. Coast Alliance, *Mission Possible, State Progress Controlling Runoff under the Coastal Nonpoint Source Control Program*, Coast Alliance, Washington, D.C., 2000, p. 66.
5. Cohn, N.C., *Voices of the Watershed, a Guide to Urban Watershed Management Planning*, Friends of the Chicago River, Chicago, 1999, p. 83.

2 Watershed Management Process

Take any road, if you don't know where you're going, and you will still get there.

— Alex Davenport

2.1 INTRODUCTION

The contemporary view is that watershed management must be viewed as a process rather than a discrete product. Professionals now have a better understanding of the need to improve coordination and implementation of various efforts on a watershed basis to protect and restore the nation's waters. Watersheds are stressed as the appropriate spatial unit for resource management agencies to assess the relative contribution of human activities to the quality and quantity of water at specific points on particular waterbodies. Typically, watershed management efforts follow a process that limits its scope and has been prescribed by either agency guidance or program regulation. While watershed management usually depends on agencies for support, no one agency can control it. No one process can solve all water resource problems; most approaches prescribed by either guidance or regulation have common elements. The common elements are combined with the necessary supporting activities to provide the basis for the recommended process that supports stakeholder involvement to solve local problems.

Water is viewed as an asset and a liability. The premise under which management agencies operate often leads to institutional conflicts and the lack of cooperation and coordination. A diverse number of federal, tribal, state, and local agencies are involved in watershed management, and each seems to have its own planning process and approach.

Currently more than 25 federal agencies have some water management-related responsibilities. The National Research Council (NRC)[1] performed an excellent analysis of the federal government and its watershed management role and responsibilities. The water resource role for most federal agencies has changed with time. Most federal agencies started with a single purpose and through legislation and amendments to existing legislation saw an expansion of their roles and missions. Unfortunately, no overall strategic plan for the management of the nation's water resources exists. This lack of a strategic plan results in program expansion that in some circumstances, due to politics, is illogical and unnecessary. The expansion of agency responsibility is usually not accompanied with the necessary resources to fulfill the new mandates. The expansion thus decreases agency effectiveness, resulting in mission fragmentation.

Start with more than 25 federal agencies, victims of mission fragmentation, add in thousands of entities interested in watershed management, ranging from international agencies to local organizations and the public, — and organizational chaos ensues. This chaos is a major obstacle to effective watershed management; disputes arise over proper focus of water-related regulations, funding, implementation, and the scale of the problem. One case in point is how agencies are dealing with the Gulf of Mexico's hypoxia problem. The Gulf of Mexico is affected by multiple sources of pollution, such as agricultural nonpoint source runoff entering streams throughout the Mississippi River basin. Burkart and James[2] identify the Midwest as the area with the largest nitrogen loss rates. Nitrogen is carried by the Mississippi River and enters the Gulf of Mexico each spring and early summer, causing a hypoxia problem known as the "dead zone." The dead zone is an offshore area affected by large algal blooms from excessive nitrogen loads in the water and characterized by a lack of dissolved oxygen in the water column. Agencies cannot deal with problems as extensive as the Gulf of Mexico's hypoxia problem as a single watershed issue, because the watershed is too large. A hypoxia task force was established to look at basin-wide problems and solutions, but most basin-wide plans or recommendations require local actions and support to be successful. Watershed efforts to reduce nitrogen can be organized and understood more clearly when addressed within the context of local watershed problems. For example, when trying to address the Gulf of Mexico hypoxia issue in the Midwest, local communities can begin by focusing on winter fish kills and groundwater contamination — issues that localize the need to address nitrogen contamination. With local interest in a problem, the community will usually be ready to act more quickly.[3]

The number of approaches to watershed-based efforts equals the number of agencies involved in water resource management. The NRC[1] outlined several federal agencies' watershed-planning efforts including the Forest Service's 10-step process, the Bureau of Reclamation's five-step process, and the Natural Resources Conservation Service's process, which covers nine steps in three phases. The U.S. Army Corps of Engineers and the U.S. Environmental Protection Agency (EPA) use principles to guide their watershed-planning efforts. All agencies have a common goal — solving problems on a geographic basis. They all share one common weakness — no specific guidance on linking the problem to the appropriate watershed scale. This can create problems when working with local partnerships. Watershed size needs to be appropriate to the land and water problems of concern. If a watershed management effort is addressing sources of pollution outside the watershed, the problem shed of interest may transcend hydrographic boundaries and involve additional stakeholders.

Even if federal agencies do not require formulaic partnership and planning models, most state and tribal agencies have established their own processes based on administrative and funding requirements and guidance. Unfortunately, this does not serve the needs of local watershed partnerships. Watershed partnerships need to work with manageably sized drainage areas but also encompass all the different, integrated, contributing areas (surface and groundwater). This enables the partnership to make measurable progress and create stronger ties between stakeholders and the waterbody they affect. Local watershed partnerships need agencies to be flexible and responsive to their circumstances by providing the appropriate tools, technical assistance, and necessary funding. Agencies need to adopt a collaborative, problem-solving, planning, and management orientation. This orientation must promote consensus-based, negotiable discussions that engender broad agreements and specific, situation-appropriate policy and management actions.

2.2 WATERSHED MANAGEMENT PROCESS

The key question for watershed partnerships is, "What is an effective process to integrate science, policy, and public participation to best manage our water resources?" A flexible process to take advantage of emerging opportunities and the unique circumstances of each watershed is needed. Below is a flexible, four-phase approach to watershed management based on a collaborative process that responds to common needs and goals. It utilizes assessments and decision processes based on a combination of biophysical, social, and economic information, plus local knowledge. An historic shortcoming has changed our understanding of watersheds and the benefits of putting integrated management into action, but a collaborative process overcomes this shortcoming. If the first two phases are properly performed, funding programs will be available to partnerships based on what they have accomplished and how they did so. Partnerships must resist the temptation of having the planning process chase funding programs. The watershed management process and written documents are a foundation, not a prescriptive tool.

Planning an entire watershed can be an overwhelming task because watersheds are complex with many interactions between humans and the environment. A dynamic, iterative process that can address a broad scope of issues; exhibits a systems orientation; and results in goals concerning healthy ecosystems, economic returns, and resource management is needed for local watershed planning. Watershed initiatives have great potential to change over time. While they should start with a narrow focus, they can become broader in scope and more inclusive of multiple goals as dictated through the management process and stakeholder involvement. For the watershed management process to be effective, people living, working, and owning or using the land in the watershed must make the right short- and long-term decisions. Their challenge is to balance short-term demands for production and services with long-term sustainability of a quality environment. The widely accepted definition from the World Commission on Environment and Development is "management that meets the needs of the present without compromising the ability of future generations to meet their needs." Utilizing this definition as a guide, watershed residents must avoid emphasizing either production or quality-of-life principles at the expense of the other; becoming involved in all phases of the watershed management process will accomplish this balancing of principles.

The "recommended process" consists of a series of four basic phases: assessment, planning, implementation, and evaluation built on stakeholder involvement and monitoring. While listed sequentially, all the phases are interconnected and are most appropriately viewed in a circular manner (as Figure 1.1 shows). Success of the monitoring effort requires clear and meaningful milestones and objectives for the plan and its implementation. Adaptive management is also required for the process to work. Adaptive management involves adjusting the management direction as new information becomes available. For example, landowners in the Double Pipe Creek RCWP area were willing to try new ideas because of the adaptive management approach. RCWP staff explained to landowners that their financial and technical assistance contracts could be modified to alter the proposed management systems, if the constructed facilities did not perform as expected.[4] Adaptive management may require the steering committee to revise a decision before moving ahead. However, adaptive management does not mean that the watershed's water-quality goals would be modified based on the lack of progress; rather the results are used to modify management policies, strategies, practices, and operation and maintenance procedures to reach the desired goals. The combination of the natural variability in the hydrologic cycle and the uncertainty associated with the off-site impact of pollution (point and nonpoint source) controls requires that the watershed management process be flexible enough to modify the implementation approach based on watershed progress and changes. Every combination of watershed characteristics, sources of pollutants, and management approaches is unique. Therefore, management efforts may not proceed exactly as planned. Meaningful, effective, and positive environmental improvements require a watershed management plan and an organization with the support and resources to carry out the plan.

The Watershed Project Management Guide provides a "recommended process" for watershed management that is based on the lessons learned through a gradual change from the top down to leading efforts locally. The Western Governors' Association's (WGA) Principles for Environmental Management in the West Policy Statement[5] contains eight principles the governors believe are critical to the success of any environmental program. Comparing the WGA's principles with this watershed management process shows that the recommended process has all the elements for a successful environmental program:

1. National standards, neighborhood solutions — assign responsibility at the right level. The recommended approach is focused on meeting water quality standards on a watershed basis with local solutions. The term "solutions" is used here to refer generally to techniques for accomplishing planning objectives.
2. Collaboration, not polarization — use collaborative processes to break down barriers and find solutions. The watershed management model is based on partnership leadership and consensus building to guide the development and implementation of the watershed management plan.
3. Reward results, not programs — move to a performance-based system. The four–phase process is a problem-solving approach based on an adaptive management philosophy responsive to performance-based achievement.
4. Science for facts, process for priorities — separate subjective choices from objective data gathering. The four-phase process uses science-based approaches, with public participation, to develop and implement local solutions to solve local problems.

5. Markets before mandates — replace command and control with economic incentives, when appropriate. A comprehensive watershed management approach recognizes the various causes of pollution and seeks a fair and equitable local solution to addressing water pollution issues.
6. Change a heart, change a nation — ensure environmental understanding. This recommended approach depends on educated stakeholders leading the process.
7. Recognition of benefits and costs — make sure environmental decision makers are fully informed. Watershed management must balance environmental protection, economic interests, and social equity in developing solutions.
8. Solutions transcend political boundaries — use appropriate geographic boundaries for environmental problems. The goal of the watershed management process is "to ensure that water and related resources are managed on a sustainable basis to provide for the environmental, social, and economic well-being of the stakeholders" and not to let political boundaries be a barrier.

2.3 THE FOUR-STEP PROCESS

This recommended process requires a partnership to manage the process to ensure stakeholder involvement, adequate monitoring, and an outreach program. A major purpose of the watershed management process is to obtain legitimate stakeholder participation early in the plan's development process. Scientists and land managers are stewards of public resources, but they need interaction with stakeholders to devise the most appropriate plan to manage the watershed to fulfill stakeholder goals. Since stakeholder opinions on watershed issues often conflict in mixed watersheds, compromise is often necessary to achieving desirable actions. Such compromise is only valuable when developed with the involvement of all the stakeholders and supported by accurate information. The more stakeholders are involved in the management process, the more effective the effort will be in developing and attaining community-based goals for public resources. Stakeholder involvement provides the foundation to obtain the reactions, consensus, and support needed in implementing the developed plan. Another key component of this process is a monitoring effort that assesses the implementation of the watershed management effort and provides the necessary data to evaluate how well the management process is working and to serve as the basis for making midcourse adjustments.

While listed sequentially, the process actually functions in a circular manner once initiated. The most logical starting place is with the watershed assessment and problem identification phase.

2.3.1 ASSESSMENT AND PROBLEM IDENTIFICATION PHASE: WHAT IS HAPPENING TO THE RESOURCE?

The assessment phase has four parts, (1) inventory/mapping; (2) analyzing information and data; (3) identifying problems (stressors and their sources) based on the analysis; and (4) determining the overall goal. Watershed assessment involves a careful analysis of all water resources in the watershed and their stressors. The geographic scope of the project is an important factor in determining the project's complexity; groundwater and surface water are fundamentally interconnected. If a partnership has difficulty combining watershed boundaries and groundwater recharge areas, the people involved need to combine surface and groundwater into a single, larger watershed boundary. Once the watershed boundaries have been delineated, identify and map the geologic features (including soils), land-use and management patterns, sensitive environmental areas, and water resources in the watershed. The water resources include the surface (wetlands, lakes, streams, and coastal) and groundwater (confined and unconfined aquifers).

In addition to the mapping of various physical features, an inventory of available data and local knowledge needs to be completed. Inventory data and information are related to the affected natural resources as well as the social and economic considerations. A complete inventory of the soil, water, animal, plant, air, and economic conditions becomes the vehicle that moves the planning committee to a position of knowledge. Local knowledge includes historical perspectives, experience with previous resource conditions and management efforts, and site-specific cultural information. The entire partnership needs to be involved in investigating the watershed, identifying the resource problems, locating stressors, and documenting conditions. Assessing and analyzing the available data and information are vital to clearly define the natural resource issues, problem sources, and critical areas. This identification provides the information needed to formulate the overall goal. As part of the assessment phase, these factors need to be integrated into a landscape-level analysis.

It is tempting to begin by "fixing" visible downstream problems without knowing the cause of the problems. Much money has been wasted treating the lower part of a watershed without addressing the real cause of the problem. The assessment process will find the causes, and the fix can occur during the planning phase. A common problem with the watershed management process is overemphasizing the assessment phase in terms of time, effort, and funding expended. This is usually attributable to the planning and technical advisory committees' underestimating the cost of data collection and the amount of data collection required and overestimating the amount of information needed in the decision-making process. Too often the majority of the planning budget is spent on the assessment/problem identification phase, leaving too few resources for adequate alternatives development and analysis. Prescreening to identify data needs, availability, and cost of collection may assist in controlling assessment cost and reducing the overall timeframe. Subsequent phases of the process should proceed within a critical period, to build on momentum among the stakeholders.

From the assessment and problem identification phase, the partnership should be able to answer the following questions:

1. What are the current environmental concerns?
2. What is the cause of these concerns?
3. What data and information are needed to complete the project?

Watershed Management Process

4. Do we have the right people to help identify and correct the problems?
5. What will be the effort's overall goals?

2.3.2 PLANNING PHASE: WHAT DO WE NEED TO DO? HOW WILL WE GET THERE?

What needs to be done in the watershed is answered here. Additional stakeholders and technical resources are usually identified and brought into the process during this phase to contribute to the refinement of watershed goals and the development of clear objectives and alternatives. The assessment and problem identification phase provides the goals, whereas the planning phase develops the road map to achieve them. The overall watershed goals usually conform to one of the three potential strategies:

1. If your resource is in good condition, maintain its existing high quality by protecting it from any further degradation.
2. Sustain the existing quality of your resource by preventing further degradation.
3. Attain an acceptable level of quality by restoring the resource from previous degradation.

Planning linked to management provides the framework for what must be done. While planning is not viewed as an action process, it is a difficult, frustrating, and time-consuming phase of watershed management. Planning and management go hand in hand in an ongoing cycle. With planning, management is more focused, meaningful, and effective. Investment in the details of preparing and implementing a management plan enhance the partnership's probability of attaining its management goals. During this phase, working with stakeholders' goals, the efforts' objectives are set and options to address the known problems are identified and selected.

Watershed management plan development usually relies on one of two approaches, "bottom up" or working "top down." The only difference in the two approaches is the means by which the overall plan is achieved. Both approaches have advantages, and the method selected needs to be based on the characteristics of the partnership and the local issues addressed. Selecting the appropriate approach will lead to wider acceptance of the watershed management effort and quicker implementation of activities.

Regardless of the approach, the partnership with stakeholder involvement has already determined the geographic scope of the effort, set the project's goals, and now must facilitate the development of a set of alternative approaches to solve the identified problems and reach the management goals and objectives. A broad range of technically feasible alternatives should be formulated that will meet the goals and objectives for addressing the identified problems, take advantage of opportunities, overcome known barriers, and prevent additional problems. As part of formulating alternatives, the stakeholders determine the feasibility of using various management options to meet the project's goals. The alternatives need to consider the projected future conditions to accommodate growth and change in the watershed. Problem identification and analysis in the assessment phase provide the linkage between water resource goals and objectives and management options. The watershed partnership uses the information regarding the water resource condition to evaluate proposed solutions in order to select the alternatives that will provide the best framework for the implementation plan. Stakeholders will select approaches based on an evaluation of the technological feasibility, on- and off-site effects, funding availability (cost sharing and other incentives), technical assistance requirements, owner/operator support, and cost to meet the overall watershed goals.

The selected alternatives define the management approaches and techniques the project utilizes to achieve its goals and objectives. A management approach is the means by which techniques are implemented. A management technique is any tool that can help a partnership manage the watershed. Techniques are generally implemented in combination with other management techniques as a system. It is important that preconceived solutions are not looking for problems, but are adequately evaluated during the alternatives analysis. Some agencies like to promote their own approaches aggressively during the planning process to meet administrative goals.

Management techniques are generally categorized by their purpose as prevention, control, or restoration. The implementation of these techniques relies on a combination of regulatory authorities and voluntary efforts dependent on the availability of technical and financial assistance and the effectiveness of the outreach effort. The application of these techniques will vary by project goals, objectives, and social setting and needs to be considered in a continuum, with greater intensity and complexity to support the restoration purpose. Prevention is associated with source controls and nonstructural techniques. Control objectives are associated with more intense methods, including source treatments and structural devices. Restoration approaches require the most intense techniques including treatment and structural techniques. As a rule of thumb, the more intense the treatment, the more costly it is to implement and maintain.

For all management techniques promoted in the plan, the necessary operation and maintenance (O&M) components must be included. As with other components in the watershed management plan, the O&M component must clearly identify responsibilities, funding sources, maintenance, and inspection schedules. This is important because it clearly identifies the long-term roles and responsibilities for organizations and agencies involved with the watershed management planning process.

Through this planning process a monitoring and evaluation component for the watershed plan is developed. In most cases initiation of the monitoring and evaluation component must occur before implementation of control and remediation approaches. Given the inherent uncertainty in management and the ongoing need for up-to-date information, monitoring and evaluation are essential in watershed management. This component normally should comprise about 10 to 15% of the management plan costs.

A good watershed management plan, developed with active stakeholder involvement, clearly identifies who, what, when, where, and how for implementation and evaluation activities. The plan must include the long-term O&M requirements and responsibilities. An important barrier to overcome during this phase is the preconceived endpoint, i.e., the watershed management plan.

Watershed Management Process

The purpose of the plan and associated level of approval will determine the amount of detail that must be included in the written plan. The steering committee has to determine the audience for the written plan. From the watershed management plan, the partnership should be able to answer the following questions:

1. What are the management goals and objectives for the project (the overall goal was set at the end of the assessment and problem identification phase)?
2. What are the priority resource areas and concerns?
3. What is proposed to be implemented: when, where, and by whom?
4. How will the partnership evaluate whether or not it has been successful?
5. How will the public and key interest groups be kept involved?

2.3.3 Implementation Phase: Making a Difference

The watershed management plan provides the framework that specifies budgets, possible funding mechanisms, timelines and milestones, action items, management strategies, and personnel assignment. Of course, successful implementation of a watershed plan requires stakeholder and landowner/operator support as well as both time and financial commitment of participating agencies and organizations. For long-term success, a watershed project needs to include active support (technical, financial, and leadership) for maintenance activities.

For restoration options, the planning committee needs to select restoration techniques and develop schedules, locations, and site-specific restoration goals (related to the overall watershed goals). Restoration practices and approaches must be tied to the structure and function of the identified water resource. The restoration techniques must not be implemented before the upstream controls are in place. Generally, the implementation phase of a watershed project involves achieving important goals.

The partnership should be able to answer the following questions:

1. What has been accomplished?
2. Are the expected improvements detected?

2.3.4 Evaluation Phase: Did We Make It?

Kondolf and Micheli[6] emphasize that despite increased requirements and commitments, evaluations have generally been neglected. Evaluating the effectiveness of the plan involves five components: monitoring, performance assessment, operation and maintenance, adaptive management, and the final report. The purpose of this phase is to evaluate the effectiveness of the watershed management plan to ensure that the implementation efforts are functioning as planned and are achieving the objectives; to identify reasons for lack of progress in plan implementation, if applicable; and to obtain information on the results of the proposed controls. Where actual results differ from those anticipated, constructive feedback needs to be provided to the planning committee. By strictly focusing on what happens, partnerships lose sight of the impact of their efforts, and they cannot publicly report the expenditure of public funds and how the watershed effort has affected society. Monitoring and assessment are used throughout the process, for different purposes, to yield different types of information that the planning committee can use in many ways. Most evaluations end when the plan has been implemented within the constraints of the budget, so the impact of the effort is not documented. The design and implementation of the evaluation effort should recognize the lag between the time a practice is applied, and adapted, and its effects on water quality and natural resource conditions. It should also be designed to recognize any effects from natural variability and past land uses and detect the effects from applied land management practices. The important step with monitoring and assessment is to answer some basic questions before any effort is undertaken: why, what, where, when, and how.

The ongoing evaluation should annually track outcome and process. Partnerships need to be familiar with the idea that at times implementation requires trial and error over time to learn how the water resource will respond to implementation efforts. Many tools exist for use during evaluations.

Evaluations that track implementation will allow for celebration through the life of the project, an opportunity for media attention, and openings for new partners to join, and they provide for the basis for any needed change of direction in the watershed management plan.

An evaluation component should answer:

1. What information/evidence is to be gathered?
2. Who will provide the information?
3. What tools will be used to gather the information?
4. Are the required tools available, or do they have to be developed?
5. How will the information be evaluated for results?
6. Who will analyze and review the results?
7. Will someone else gather and analyze the data?
8. What are the overall evaluation guidelines and design?
9. How are matters of privacy dealt with?
10. How will evaluation results be used?

If your evaluation does not answer the "right" questions, it may lead to a sense of nonaccomplishment for the project and the development and acceptance of inadequate and inappropriate management approaches.

From the evaluation activities, the partnership should be able to answer the following questions:

1. What worked?
2. What did not work?
3. Are there sites or critical areas that need additional treatment?
4. Is long-term maintenance occurring?

Answering these questions allows the partnership to determine its impact and if anything needs to be changed. Effective evaluation provides the foundation for the future.

2.4 SUMMARY

The process and the necessary support components provide mechanisms for solving watershed-level problems. Local leadership must define what needs to be done based on an understanding of what is possible within the framework of the physical and chemical watershed processes.

REFERENCES

1. National Research Council (NRC), *New Strategies for America's Watershed*, National Academy Press, Washington, D.C., 1999.
2. Burkart, M.R. and James, D.E., Agricultural-nitrogen contributions in the Gulf of Mexico, *J. Environ. Qual.*, 28, 850, 1999.
3. Davenport, T.E. and Wilson, D., Local Solutions to Regional Problems: The Watershed Approach, presented at Earth Day Symp., St. Louis, April 26, 2001.
4. Schaeffer, E.A., Farmer participation in the Double Pipe Creek, Maryland, Rural Clean Water Program Project, in *Proc. The Rural Clean Water Program Symposium, 10 Years of Controlling Agricultural Nonpoint Source Pollution: The RCWP Experience*, EPA/625/R-92/006, USEPA, Washington, D.C., 1992, p. 265.
5. Western Governors' Association (WGA), Principles for Environmental Management in the West, Policy Resolution 99–013, WGA, Denver, 1999, p. 6.
6. Kondolf, G.M. and Micheli, E.R., Evaluating stream restoration projects, *J. Environ. Man.*, 19(1), 1–15, 1995.

3 Watershed Processes

Water is the most critical resource issue of our lifetime and our children's lifetime. The health of our waters is the principal measure of how we live on the land.

— Luna Leopold

3.1 INTRODUCTION

Watershed planning uses watersheds and subwatersheds as the biophysical basis for planning and management. The classic definition of a watershed, "an area of land that drains to a common point," explains the ultimate purpose and role of a watershed. They are based on using the hydrologic cycle as the pathway that integrates physical, chemical, and biological processes of the watershed. Watersheds give a fundamental unit with real boundaries of ecological significance that in turn provide a quantitative reference to examine issues such as ecological stress, the cumulative effects of land-use development, and other matters. When using watersheds as a basis for management, decision makers must be aware of many important hydrologic considerations. Because so much has been written about these considerations, this chapter presents only a brief overview. It is not comprehensive but rather a guide outlining considerations for developing and implementing a watershed management plan. Additional resource material for further information is provided at the end of the chapter.

Four primary factors affect the quality and function of resources in the watershed: water regime, flow regime, habitat, and energy source (Figure 3.1). These factors influence the overall health and viability of the watershed's water resources. Changes in any factor can change the watershed and the water resource under consideration. Managing the watershed is managing this multitude of factors to secure the integrity and balance of the whole system and the highest-quality water resource. The water-quality factor includes the physical and chemical characteristics of the water. Flow regimes encompass water volume, temporal distribution of floods and low flows, and water velocity. Habitat relates to the aquatic and near-shore substrates. Energy sources include the organic material entering a water resource, the process of photosynthesis, and the seasonal pattern of available energy from sunlight.

Water quality is impaired by land uses that contribute pollutants to runoff or groundwater. The following conditions adversely affect water quality: elevated nutrient inputs, excess sediment, pathogens, toxic substances, elevated levels of organic matter, low dissolved oxygen, and increases in water temperature. These conditions

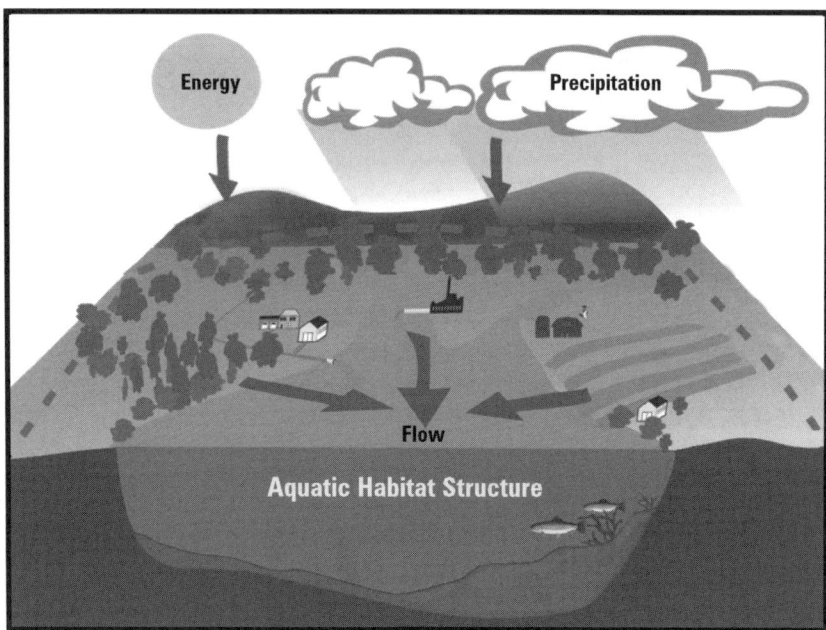

FIGURE 3.1 Factors affecting watershed health. (Courtesy of Conservation Technology Information Center (CTIC), West Lafayette, IN.)

can cause human health and aquatic life support impairment. For example, the surface-water impacts of phosphorus loadings have little to do with human health concerns and relate instead to aquatic life and aesthetic and economic concerns. Nutrient sources can be the most widespread and difficult nonpoint source to identify and quantify. Watershed sources include commercial fertilizers, animal waste, soil reserves, and atmospheric deposition. Stream- and lake-bottom sediments can release stored nutrients into the water column.

The physical characteristics of the watershed define the structure and function of the aquatic and near-shore habitats. Changes in habitat structure and function can be caused by modification in the morphology, condition of banks and upland areas, presence or absence of vegetation and debris, and substrate quality. Changes in the habitat affect the composition and abundance of plant and animal species and the stream flow. Flow regimes are determined by the interaction of the watershed's physical characteristics, human interaction, and climate. Perennial water resources have two common conditions: low and high flow. High flow is also known as wet-weather flow. Changes in flow regimes can cause channel instability, destroy aquatic habitats, and lower the water table. Research shows that hydrologic stability is a factor of the watershed-to-wetland ratio. Results of the Illinois State Water Survey indicate in order to maintain hydrologic stability the wetland-to-watershed ratio should be at least 12%.[1] Energy sources within the watershed depend on two processes: photosynthesis and metabolism. Changes in the energy source can disturb the food web.

TABLE 3.1
The Number of Watersheds by Scale/Classification in the U.S.

Scale	Classification (Digits)	Number	Example
Region	2	21	Upper Mississippi River
Subregion	4	222	Illinois River
Basin (accounting unit)	6	352	Upper Illinois River
Watershed (cataloging unit)	8	2,150	Lower Fox River, Illinois (more than 448,000 ac)
Subwatershed	11	Over 7,000	Fall Creek, Indiana (250,000–448,000 ac)
Microwatershed	14	Unknown	Pittsfield City Lake, Illinois (about 35,000 acres)

Addressing the symptoms of a watershed problem without correcting the source and cause of the problem is not only short-sighted, but also foolhardy. The four factors discussed above act independently and interdependently to influence the overall condition and viability of the water resource. Impacts or stress on a single factor can cause a change in another. Watersheds within an ecoregion share characteristics that establish a common basis for management goals and objectives. Soils, terrain, climate, and vegetation are similar for watersheds within the same ecoregion.

3.2 WATERSHED UNITS

The four-step process is based on "managing by watershed" and recommends the use of U.S. Geological Survey hydrologic units. A national hierarchical framework of hydrologic unit codes was developed by the USDI-Geological Survey in cooperation with the U.S. Water Resource Council.[2] It provides a common national framework for delineating watersheds and their boundaries at a number of different scales. Table 3.1 shows the breakdown of the number of watersheds by scale in the U.S.

The numbering of watersheds is consecutive from upstream to downstream. The first two digits indicate the main river basin. The third digit may indicate all or a portion of that basin. In this way the third digit can be changed to designate, for example, two parts of a watershed on either side of a state line. The larger the hydrologic unit code number, the smaller the watershed. These watersheds refer only to surface-water resources and do not necessarily reflect groundwater contribution. The advantages of using the hydrologic unit codes are as follows:

The hydrologic code attached to a specific watershed is unique.
The code provides a common language for different organizations and agencies to use. If a code has been assigned, then there is agreement over the watershed's boundaries.
Having watersheds delineated on published maps assists the public in understanding how landscapes function, where water-quality problems may be

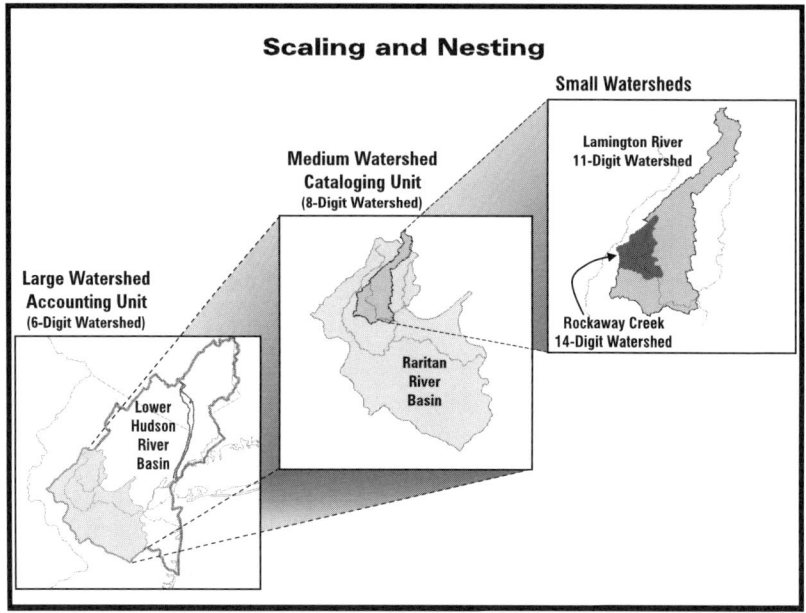

FIGURE 3.2 The scaling and nesting of watersheds. (Courtesy of CTIC, West Lafayette, IN.)

addressed, and geographically who needs to be involved in the planning process.

For the watershed management process, partnerships need to adopt watersheds and subwatersheds as the fundamental analytical and management units. While watersheds are nested and hierarchical elements of river systems, the individual units have definable boundaries (see Figure 3.2).

However, a watershed's drainage area consists of two components: groundwater and surface water. Groundwater and surface water are interconnected. Often it is difficult to separate the two because they "feed" each other. In many xeric regions of the U.S. where watersheds can be defined and influent streams predominate, topographic watersheds do not encompass the same integrating processes as in mesic and hydric areas. About 40% of river flow nationwide depends on groundwater. The surface drainage basin, or the land area from which all surface water flows, drains toward a surface-waterbody at a lower elevation. The groundwater drainage basin, or the land area and associated subsurface through which groundwater drains, drains to a surface-waterbody at a lower elevation. Together they make up the watershed for the downstream surface-waterbody.

3.3 DRAINAGE NETWORK

Streams join to form a branching network. This network empties into or can be an outlet for a lake or estuary. A numerical system exists that describes this network; it is referred to as stream order. A first-order stream has no tributaries; when two

TABLE 3.2
Estimated Number and Length of River Channels in the U.S.

Strahler Stream Order	Number of Streams	Mean Length (mi)	Mean Drainage Area (mi^2) — Cumulative Drainage Area
1	1,570,000	1	1
2	350,000	2.3	4.7
3	80,000	5.3	23
4	18,000	12	109
5	4,200	28	518

Source: Modified from Schueler, 1995, Environmental Land Planning Series: Site Planning for Urban Stream Protection, Metropolitan Council of Governments and Center for Watershed Protection, Washington, D.C., p. 89.

first-order streams join, they create a second-order stream. When two second-order streams join, a third-order stream is created, and so on until tenth-ordered streams are formed. The smallest-order streams are the most numerous due to the geometric arrangement of stream networks. Within a given watershed, stream order correlates well with other watershed parameters, such as drainage area and channel length. The Mississippi River has been classified as a 10th-order stream. Table 3.2 gives the estimated number and length of channel for stream orders up to order 5 When a drainage area is more than 500 mi^2, it is too large to address as a single watershed management unit. As the stream order increases, many changes may occur in the physical, chemical, and biological characteristics of the system. Figure 3.3 shows stream ordering using the Strahler method.

The term *time of concentration* is used to describe the time needed for runoff to move from the hydraulically most distant part of a watershed to a point of interest. Alterations to the land surface, drainage network, and vegetation can have a dramatic effect on the time of concentration. The pattern of stream channel system within a watershed can be described by drainage density (total length of drainage system /area of the watershed). Drainage density reflects the response of the watershed to precipitation, in terms of the natural function to move water and constituents from a place of higher elevation to a place of lower elevation.

3.4 WATER CYCLE

The topography and types of soils are a determining factor in land-use patterns. The water (hydrologic) cycle describes the continuum of the transfers of water from precipitation to surface water and groundwater, to storage and runoff, and to the eventual return to the atmosphere by transpiration and evaporation. The hydrologic cycle begins with precipitation in the form of rain, snow, sleet, or hail falling on the land surface. As precipitation falls, some may evaporate directly into the atmosphere from bodies of water, and a portion may be intercepted by vegetation. The remainder reaches the ground where it can enter the soil by a process called *infiltration*. Some infiltrating water remains near the soil surface and evaporates into the atmosphere. Another portion is extracted by plant roots and transported to leaves where it is lost

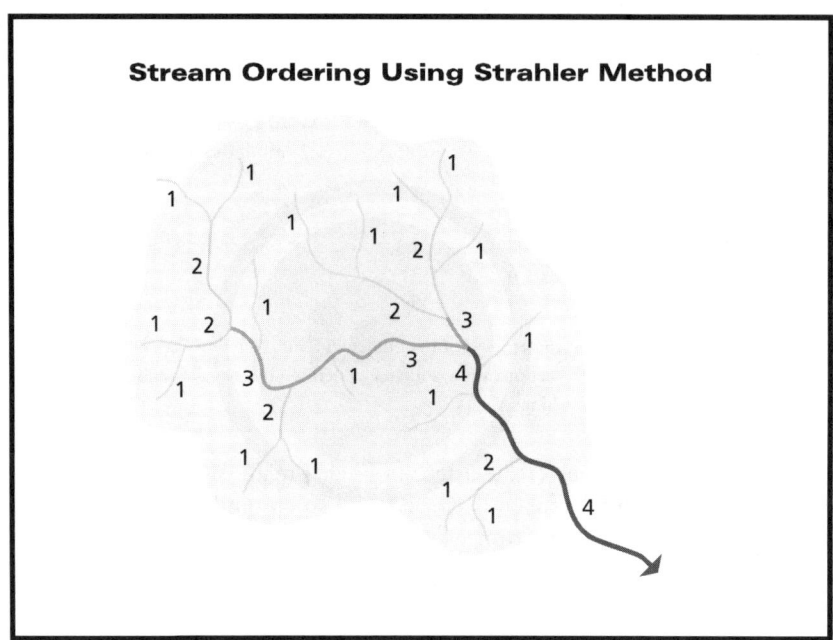

FIGURE 3.3 Stream ordering using Strahler method. (Courtesy of CTIC, West Lafayette, IN.)

to the atmosphere as vapor. This process is called *transpiration*. Water will infiltrate into the soil as long as the potential rate of infiltration exceeds the rate of precipitation. When the precipitation rate exceeds the infiltration rate, excess water builds on the soil surface and moves by overland flow called surface runoff. This surface runoff can pick up and carry pollutants from the land surface. If surface runoff is excessive or concentrated, soil erosion can occur. Still another portion of water that enters the soil can move vertically or laterally out of the plant root zone. Significant lateral movement of water through the soil is called interflow. Downward movement of water through the soil is referred to as *percolation*. Percolating water eventually makes its way to a saturated zone, where all spaces between rock and soil particles are filled with water. The water filling the spaces between soil particles and rock in the saturated zone is groundwater. Figure 3.4 shows this.

Whenever humans make changes on the land surface or to its vegetation, some aspect of the water cycle is altered, with some concomitant effect on runoff, and stream flow occurs. These effects may include the amount, timing, and location of water reaching the drainage network. The continuity within the drainage network leads to the off-site impacts of the alteration being manifested at distant locations. Watershed management becomes a process to minimize the off-site impacts of this alteration at various points in the landscape.

3.4.1 Precipitation

The type of precipitation that occurs is generally a factor of humidity and air temperature. Geographic location relative to large waterbodies and topographic relief

Watershed Processes

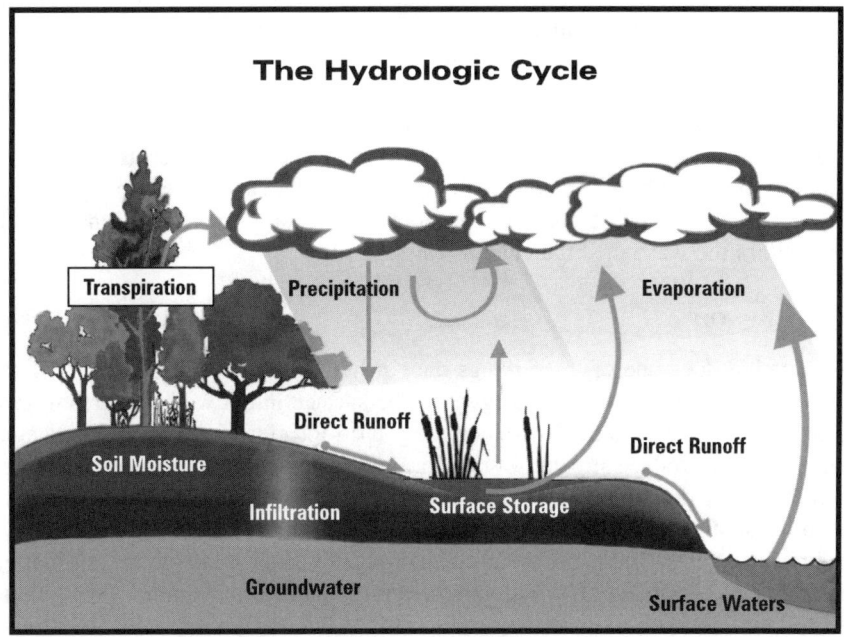

FIGURE 3.4 The hydrologic cycle. (Courtesy of CTIC, West Lafayette, IN.)

affects the frequency and type of precipitation. For example, snowfalls occur more frequently at high elevation and mid-latitude areas with cold seasonal temperatures. Approximately two thirds of the precipitation over the U.S. evaporates to the atmosphere. This occurs because of two processes: interception and transpiration. A portion of precipitation never reaches the ground because vegetation and other natural and constructed surfaces intercept it. Transpiration is the diffusion of water vapor from plant leaves to atmosphere. Evaporation is the diffusion of moisture from ground and surface to the atmosphere. The amount of precipitation for a location drives transpiration and evaporation. However, soil type and vegetation-rooting characteristics also play roles in determining the actual rate of evapotranspiration.

Precipitation analysis is essential to making good decisions about managing watershed runoff. Precipitation has a time and spatial distribution. Time distribution or duration refers to the amount of precipitation received in any period of time. Spatial distribution refers to the amount of precipitation received from region to region and ranges from 50 to 60 in. in the southeast to 30 to 40 in. in the north. Precipitation analysis provides a method to calculate the annual rainfall as well as categories of rainfall patterns given its variable nature. A number of national publications and sources of rainfall data and information are available: the EPA data set is divided into 12 rain zones to represent the U.S., and the NRCS divides the U.S. into 4 zones. The National Weather Service of the U.S. Department of Commerce is the best source of consistent precipitation nationwide. These national sources should be considered default data sets if a representative data set is not available for the watershed of concern. Both total amount and intensity of rainfall are important influences on nonpoint source pollution. Intensity refers to the rate at which precipitation falls. For example, in general, a short-duration, high-intensity rainfall will cause more runoff than a long-duration, low-intensity rainfall of the same amount. A number of structural management practices have design related to precipitation duration (hr) and frequency (yr). The two most common design storms are the 10-yr, 24-hr storm and 25-yr, 24-hr storm. Local rainfall curves, called IDF (intensity, duration, frequency) curves, are usually available from state water supply or flood control agencies. Local IDF curves are more representative of annual runoff conditions within the watershed than annual rainfall.

3.4.2 Runoff

Precipitation does one of three things once it hits the ground. It can return to the atmosphere, move into the soil, or run off the surface into a wetland, stream, lake, or other waterbody. "Runoff" is a summary term to refer to the various processes that ultimately produce stream flow. According to Dunne and Leopold,[3] "The paths taken by water determine many of the characteristics of landscape, the generation of storm runoff, the uses to which it may be put, and the strategies required for wise land management." The pathways precipitation takes after it falls to the earth affect many aspects of stream flow including quantity, quality, and timing. Precipitation can follow three possible paths from the time it hits the ground until its ultimate discharge to a surface water:

1. The water may flow over the land surface without infiltrating.
2. The water may infiltrate to groundwater, then flow toward and eventually discharge into surface water.
3. The water may infiltrate and move as interflow.

Surface runoff is generally produced by one of two processes. The first is known as *Hortonian runoff* and occurs when rainfall intensity exceeds the infiltration capacity of the soil. Excess water collects on the soil surface and travels downslope as runoff processes including geology, climate, topography, soil characteristics, and vegetation factors affect it. It is common where soil disturbance limits infiltration into the soil. Typical soil disturbances are tillage, construction, and heavy traffic. Hortonian overland flow is the dominant storm runoff process in arid and semiarid regions. The second process, known as the *variable source area process*, is dominant in most humid regions where infiltration is not a limiting factor. This process occurs when the soil becomes saturated and can easily occur even on areas with heavy residue cover. Soil saturation is common in areas where the soil is shallow to bedrock or an impermeable layer, and it depends on the complex topography in a specific location. The variable source concept assumes that the source of water for stream flow varies over time. Just as the source of water varies, so does the source of constituents, mainly particulate pollutants, that runoff carries. The extent of land's contribution to stream flow is called the "effective watershed." Areas that do not contribute water to the stream flow must be properly mapped to obtain an accurate estimate of effective drainage area. Otherwise the amount of runoff and pollutant

Watershed Processes

loading will be under- or overestimated. An effective watershed area highly depends on climatic conditions, and it can vary year to year.

Surface runoff is the main mechanism for transporting nonpoint source pollutants from the land to surface water. Runoff transports pollutants in dissolved forms and in forms adsorbed to sediment. Soils near stream channels usually have higher antecedent moisture condition than soils farther away from the channels. As rain falls, the areas saturated first are usually closest to the channels, increasing over time as saturated soils expand laterally away from the channel. Once soil becomes saturated, the water moves across the land surface and the process of runoff begins. The detachment and transport capacity of runoff depend on the velocity and depth of flow. Topographic factors such as slope and surface depressions determine flow velocity and surface storage. The velocity and depth of flow change with time and space as runoff flows over the land surface. As runoff occurs, the process of pollutant detachment and transport occurs, and as the source of the water changes, the production of pollutants also changes.

3.4.3 Stream Flow

Stream discharge is a critical explanatory variable for quantifying the mass load of pollutant delivered from a watershed. Stream flow can be divided into two basic components. One is base flow, precipitation that percolates to the groundwater and moves through substrate before reaching the channel. It sustains stream flow during periods of little or no precipitation. The second is storm flow, precipitation that reaches the channel over a short time through overland or underground routes. Stream flow at any one time might consist of water from one or both sources. Total precipitation, antecedent precipitation, soil moisture, ground cover, and storm characteristics influence the quantity of stream flow produced at any rainfall event.

To adequately characterize the quality of a waterbody, information and data must be available for a range of hydrologic conditions to estimate pollutant loadings. Flows range from no flow to flood flows over a variety of time scales. Seasonal variations of stream flow are more predictable. Because management and control practice design work requires using historical information (period of record) as a basis for designing for the future, flow information is usually presented in a probability format. Two probability formats most useful for planning and designing watershed management are flow duration and flow frequency.

Flow duration is the probability a given stream flow was equaled or exceeded over a period of time. Duration of flow (period of time which flows occurs) can also be used to classify streams. Perennial streams flow almost throughout the year (at least 90% of the time). Intermittent streams flow generally only during the wet season, about 50% of the time or less. Ephemeral flow generally occurs during and for a short time after a storm event. Flow frequency is the probability a given stream flow will be exceeded in a year.

3.4.4 Groundwater

Precipitation that is not intercepted or flows as surface runoff moves into the soil. Once there it remains in the upper layer or moves downward to recharge groundwater.

Geology controls percolation and groundwater seepage and the mineral content in surface and groundwater.

Groundwater is often called the hidden resource in the U.S. Approximately 50% of the population's drinking water is supplied by groundwater; it makes up 95% of the rural population's drinking-water supply. Over half the irrigated cropland uses groundwater, and it fulfills approximately one third of industrial water needs. There are two threats to groundwater: quantity and quality.

The quantity of groundwater demanded to meet human needs is increasing. The three typical threats are overdraft, drawdown, and subsidence. Overdraft occurs when groundwater is withdrawn faster than recharge can replace it and can be a permanent loss. Drawdown is temporarily lowering the water tables, which corrects itself when the groundwater supply is replenished. Subsidence is a result of overpumping; the water table declines, water pressure is reduced, and the soil compacts. Subsidence reduces storage capacity, and the land can sink from a few inches to several feet.

The threats to quality are inorganic components, pathogens, and organic components. They can come from three pollution sources: point, nonpoint, and other sources. Point and other sources are usually regulated, whereas nonpoint sources are not. The point sources inject wastewater and other pollutants into groundwater in accordance with discharge or injection permits. Examples of the other sources are underground storage tanks and septic systems. Infiltration transport dissolves contaminants into groundwater. The rural sources of contaminant are fertilizers, livestock wastes, and land application of manure, sludge, and wastewater. In urban areas a number of businesses can threaten groundwater with a wide variety of potentially contaminating substances.

3.5 GEOMORPHOLOGIC PROCESSES

Geomorphology is the study of surface forms of the Earth and the processes that developed those forms. Three primary geomorphic processes are involved with flowing water: erosion, the detachment of soil particles; sediment transport, the movement of eroded soil particles in flowing water; and accelerated sediment supply. The last causes accelerated bank erosion rates, degradation, aggradation from channel disturbance, and stream-flow changes. Sediment budget increases or decreases can lead to channel change. Sediment deposition, characterized as the settling of eroded soil particles to the bottom of a waterbody or as particles left behind as water leaves, can accelerate channel instability. Sediment deposition can be transitory, as in a stream channel from one storm to another, or more or less permanent, as in a larger reservoir.

Soil erosion includes the processes of detachment of soil particles from the soil mass and subsequent transport and deposition of sediment particles on land surfaces.

> Soil erosion has several major consequences: (1) the original sites of the eroded material are degraded; therefore potential productivity is lessened; (2) the sites of deposition of the soil particles are altered physically, chemically, and hydrologically; and (3) the chemical and physical nature of the transporting water system can be drastically affected.[4]

Watershed Processes 31

According to the USDA-SCS,[5] erosion is the source of 99% of the total suspended solids loads in U.S. waterways. Soil material detached during a rainfall runoff or snow-melt runoff event is considered sediment. The source of sediment pollutant is difficult to identify because the sources are often widespread. For instance, sediment can originate from cropland, ditches, gullies, roads, forests, and stream banks. Sediment can also reenter the water column as a result of in-stream scouring and resuspension in lakes due to wind or boating activity. The concentration of sediment from existing urban areas is generally lower than that for rural areas. However, the total load is similar. This is because urban areas generate more runoff on a per-unit basis; the higher runoff volume with a lower concentration from the urban area equals or exceeds the load from rural areas that have lower runoff volumes with higher concentrations of sediment. A sediment survey and sediment budget are needed to identify watershed sediment sources, determine sediment delivery, quantify the relative contribution of each source, and identify possible management opportunities.

3.6 ASPECTS FOR MANAGEMENT

Watersheds are highly interactive systems. They integrate the surface and groundwater flow of water up-gradient from the waterbody under consideration and allow specific accountings to be made of factors such as point and nonpoint source pollutants, whose transport is associated with the movement of water. Watersheds are essential for these purposes. The condition of a waterbody is the result of the complex interaction of different physical, chemical, and biological factors. The physical and chemical factors in turn support a community of biological organisms unique to the watershed. Rainfall cycles, watershed characteristics, and waterbody and human interaction all contribute to this condition. Understanding the fundamental ecosystem processes is critical to effectively managing a watershed.

Land-use changes in the watershed impact the biological, chemical, and physical processes that correspondingly alter the water resource's systems. These systems normally function within natural ranges of water quality and quantity, termed "dynamic equilibrium." For example, a stream will attempt to transport the sediment delivered to it with available stream flow. The balance in sediment input/output is central to the equilibrium of the river channel. When the system is forced to function outside its "normal" range, the processes will make adjustments in the structure and function of the system and a new equilibrium will eventually develop. The stream network will experience either aggradation or degradation. Aggradation is raising the streambed elevation, increasing the width/depth ratio, and correspondingly decreasing the channel capacity. Degradation is lowering the local base level or the stream bank's abandoning floodplains, which lowers the water table and increases the bank height, which adds to bank erosion and leads to long-term instability. Without intervention the timeframe for development of the new equilibrium can be lengthy, and the changes to the processes necessary to achieve this new balance can be significant. Interrelated stream system variables include the size of the watershed; the amount and size of sediment transported in the stream system; the stream channel shape, slope, and size; and amounts and frequency of stream discharges. A stream

TABLE 3.3
Percentage of Samples, by Land-Use Category and Media Type, with One or More Pesticides

Land-Use Category	Fish	Streams	Groundwater
Agricultural	85	92	59[1]
Urban	100	99	49[1]
Mixed land use	96	100	33[2]

[1] Shallow groundwater.
[2] Major aquifers.

Source: USGS, 1999. The Quality of our Nation's Waters — Nutrients and Pesticides, U.S. Geological Survey circular 1225, Reston, VA, p. 58.

in which these variables are in balance is considered in equilibrium and sometimes described as a graded stream. Variables change among stream reaches and over time; the dynamic equilibrium of a stream represents the average condition during its recent history. Under conditions of dynamic equilibrium, the sediment loads entering a stream reach are roughly equal to those leaving it. Fluvial geomorphology is specifically concerned with the influences of water and rivers on the erosional cycle of land deposition and degradation over time.

3.7 LAND USE

The most significant remaining impact to the health of our nations' water resources today is from stormwater runoff, or nonpoint source pollution. Although most of the pollutants are transported in surface runoff, some may enter waterbodies through atmospheric deposition, from direct application, or from subsurface or shallow groundwater flow. During storm events, pollutants are washed into waterbodies from farms, backyards, streets, and construction sites. The old urban solution to rainwater of moving water downstream as quickly as possible has resulted in the degradation of thousands of miles of streams. Exacerbating this pollution are hydrologic impacts caused by new upstream developments, which increase flooding and associated environmental impacts of erosion, siltation, and channel enlargement. The resultant erosion of stream banks decreases the habitat of aquatic life essential to healthy waterbodies.

Usually a water resource in a predominantly agricultural or an urban watershed has a lower potential habitat condition than one in a forested watershed. The overall ecological condition of the resource is limited by the present and potential habitat conditions. Patterns of occurrence provide new perspectives about key sources of contaminants. For example, urban areas, which cover less than 3% of the land in the continental U.S., traditionally have not been recognized as important contributors to nonpoint source contamination, especially when compared with agricultural land, which covers more than 60% of the Unites States. Table 3.3 indicates that land use is not a good indicator of pesticide contamination in the U.S.

3.7.1 Urban

The impacts associated with the increases in imperviousness due to urbanization are decreases in the time of concentration of overland flow and reductions in long-term upland sediment yield. For a given rainfall event, urbanization yields more runoff and delivers this runoff to the channel more quickly than for other existing land uses. The runoff includes less sediment but more heavy metals, fecal coliforms, and other pollutants. Pollutants that tend to be specific to urban areas are trace metals (lead, copper, zinc) and other toxic materials such as oil and grease (hydrocarbons). Traditionally in urban areas, the wetlands have been lost (to either filling or excessive flow), lakes are impacted by nonpoint source pollution and modified drainage, and streams are out of balance due to excessive sediment deposition or accelerated erosion. Sediment from urban areas contains flakes of metals from rusting cars, particles from automobile exhaust, bits of tires and brake linings, particles from roof and gutter wear and tear, pieces of pavement, and smokestack emissions deposited from the atmosphere.

3.7.2 Agriculture

Like other land uses, agriculture can affect both surface and groundwater. Agriculture can contribute point and nonpoint source pollution. Point source pollution is usually characterized as a release of pollutants through a pipe or manmade conveyance. Examples of agricultural point sources of pollution are runoff from a feedlot pen or overflows from a hog lagoon to a stream. Since a point source is usually concentrated at its discharge point, it is easy to resolve. Nonpoint source pollution does not originate from one location. Below are examples of agricultural nonpoint source pollution. The examples are to highlight that nonpoint source pollution control tends to be very difficult and usually requires a change in management practices.

The Idaho RCWP Project: Rock Creek's streambed quality and trout reproductive capacity were reduced by siltation, and transparency was reduced by high suspended sediment concentrations. At the onset of the project, agricultural sources were identified as the primary cause of reduced streambed quality. Further analysis showed stream-bank erosion was also a major contributor of sediment load. The influx of sediment from stream-bank erosion made it difficult to document the effectiveness of cropland BMPs. From the project estimates, the sediment contributions from the two major sources, stream-bank erosion and irrigation return flow, were similar in magnitude to when the project began. In contrast, from 1987 to 1990, monitoring indicated that stream-bank erosion contributed from two to over five times the amount of sediment added from cropland in the subbasins during the May through August irrigation season.[6]

In the Illinois RCWP project, turbidity, siltation, and nutrients were thought to threaten Silver Lake, the water supply for the city of Highland. Sediment survey results showed that siltation was low, which meant there was little threat of rapid loss of lake storage capacity. Analysis of lake turbidity, which increased the cost of treatment, was due primarily to suspended soil particles. Monitoring demonstrated that the loading of fine-particle natric soils and their suspension from lake sediments were the primary factors causing lake turbidity. To target pollutant sources, the project placed special emphasis on keeping natric soils in place and reducing their delivery into the lake.[6]

In Illinois between 1993 and 1995, Vermillion River watershed farmers applied an average of 176 pounds of nitrogen per acre (lb/ac) to corn fields that had a recommended rate of 153 lb/ac annually. The pattern is similar for the Big Ditch Watershed, monitored by the Illinois State Water Survey. Over three seasons, farmers in the watershed applied nitrogen on average of 53 lb/ac per year in excess of the recommended rate to corn acres, which resulted in an excess of 2 million pounds of nitrogen being applied.[7]

For the Vermont RCWP project, the issue was to address significant phosphorus loading to St. Albans Bay. Phosphorus originated from a point source, agricultural operations, bay sediment, and a wetland adjoining the bay. Project area soils also contributed part of the total phosphorus load. A budget of all major phosphorus sources was needed to determine the potential for reducing lake or bay phosphorus levels and to identify opportunities for management.[6]

The Minnesota RCWP project found high nitrate levels in project-area domestic wells. Sources of nutrients included animal operations and cropland. The topography is karst limestone with extensive sinkhole formations. Sinkholes were thought to be a primary source of conveyance to groundwater until lysimeter studies showed rapid leaching of nitrate from fertilized cropland to be the major source. Further study indicated that cropland should have been targeted for treatment and sinkholes given a lower priority.[6]

3.7.3 Atmospheric Deposition

The atmosphere can also be a major source of contaminants. Atmospheric deposition can enter the water from the atmosphere either as precipitation or in a dry form. This type of nonpoint source pollution is particularly problematic in lakes throughout the northern and northeastern U.S. and Canada, as well as estuaries along the Atlantic and Gulf coasts. In many cases, atmospherically deposited pollutants have traveled substantial distances by wind currents. Numerous studies indicate that 80% of the toxic chemicals entering Lake Superior result from atmospheric deposition rather than water discharges. The USGS[8] reported that almost every pesticide investigated has been detected in air, rain, snow, or fog throughout the country at different times of the year. Fisher and Oppenheimer[9] estimated that as much as 25% of the nitrogen entering the Chesapeake Bay comes from the atmosphere. Along the Gulf Coast in Tampa Bay, 28% of total nitrogen loading enters bay waters directly through dry fall or precipitation. "Acid precipitation" is the term used to refer specifically to wet atmospheric deposition. Major sources include emissions from the combustion of fossil fuels used for transportation and the generation of electrical power. Other sources of atmospheric deposition are:

Incinerators, automobiles, and other mobile and stationary sources' release to the air
Metals, organic chemicals, and nitrogen
Sulphur compounds, which contribute to acid deposition

For more information on hydrology and geomorphology, see Dunne, T. and Leopold, L.B., *Water in Environmental Planning*, W.H. Freeman and Co., New York, 1978.

Federal Interagency Stream Restoration Working Group (FISRWG), Stream Corridor Restoration: Principles, Processes, and Practices, at www.usda.gov/stream_restoration, 1998.

For more information on groundwater, see *Citizen's Guide to Groundwater Protection*, EPA 440/6–90–004, USEPA Office of Water, Washington, D.C., 1990.

REFERENCES

1. The Wetlands Initiative (TWI), ADID Wetland No. 55, Protection Plan and Watershed Development Guidance Manual, TWI, Chicago, 1997.
2. Seaber, P.R., Kapinos, F.P., and Knapp, G.L., Hydrologic Unit Maps, U.S. Geological Survey Water-Supply Paper 2294, USDI, Denver, 1987.
3. Dunne, T. and Leopold,L.B., *Water in Environmental Planning*, W.H. Freeman and Co., New York, 1978, p. 818.
4. Davenport, T.E., Soil Erosion and Sediment Transport Dynamics in the Blue Creek Watershed, Pike County, Illinois, EPA/WPC/83–004, Illinois EPA, Springfield, IL, 1983.
5. USDA–Soil Conservation Service (SCS), *The Second RCA Appraisal: Analysis of Conditions and Trends*, U.S. Govt. Printing Office, Washington, D.C., 1989.
6. USEPA, Evaluation of Experimental Rural Clean Water Program, EPA-841-R-93–005. USEPA, Office of Water, Washington, D.C., 1993.
7. Agri-View, *Nitrate in Water Is a Matter of Rates and Timing*, Agri-View, Madison, WI, 1999.
8. USGS, Pesticides in the Atmosphere, U.S. Geological Survey Fact Sheet FS-152–95, USDI, Sacramento, CA, 1995.
9. Fisher, D.C. and Oppenheimer, M., Atmospheric nitrogen deposition and the Chesapeake Bay Estuary, *Ambio* 20(3–4), 102, 1991.

Table 3.2. Schueler, T., Environmental Land Planning Series: Site Planning for Urban Stream Protection, Metropolitan Council of Governments and Center for Watershed Protection, Washington, D.C., p. 89, 1995.

Table 3.3. USGS, The Quality of our Nation's Waters — Nutrients and Pesticides, U.S. Geological Survey circular 1225, Reston, VA, p. 58, 1999.

4 Partnership Development and Operation

Never doubt that a small group of thoughtful, committed citizens can change the world. Indeed, it's the only thing that ever has.

— Margaret Mead

4.1 INTRODUCTION

Building and maintaining partnerships is key to effective watershed management. Historically, the best watershed management efforts have been made through citizens' organizations focused on identifying and solving local watershed problems. It has been emphasized in many forums that effective watershed management is not possible without partnerships and public participation and support. The history of grassroots support for the overall Clean Lakes Program can be attributed to local involvement and management over the direction and scope of the individual projects rather than traditional USACE water resource projects and USDA Small Watershed Program projects. Unfortunately, many government-sponsored watershed efforts have not always embraced this philosophy. Meaningful, effective, and positive environmental improvements require a watershed management plan and an organization with the support and resources to carry out the plan. Through a watershed partnership, different individuals and organizations come together to address concerns and interests. This chapter focuses on the organizational aspects of successful watershed management efforts.

State and federal agencies are understaffed and underfunded to adequately protect this nation's water resources. They need to find a mechanism to do more with less. Watershed partnerships are one such mechanism. Watershed partnerships bridge the capabilities, assets, and resources of multiple agencies, organizations, and individuals. Developing and implementing a watershed plan through a partnership creates local ownership and consensus for action. In addition to consensus building, partnerships often result in:

More efficient use of financial resources
More creative and acceptable ways to manage and protect natural resources
A community commitment to natural resources overall

A partnership's true value is not in being able to control something, but in enabling others to do things. A formula for successful partnership can be clarity and commitment to outcome, outputs, and operating procedures. Partners need to have a complete understanding of and commitment to these three elements of their work.

What is a partnership? A partnership is an association of two or more people who come together for the purpose of carrying out an activity. Other terms such as "alliances," "coalitions," and "groups" have been used instead of "partnership." Whatever term is used to describe watershed organizations implies place-based focus, multiple parties, and a reliance on science to support decision making. So it is not important what the organization is called as long it fulfills the partnership role. What the partnership is called and how it is organized are entirely up to the group that forms to address the issues of concern. It is generally thought that there are two major types of partnerships: operational (partnerships focus on coordination of issues and are jurisdictional) and planning (partnerships are related to a specific geographic area of concern). For the purpose of this book, partnership is defined as an association of persons, organizations, and agencies joined in a geographically based undertaking as shareholders or partakers to address a problem or issue.

4.2 BUILDING PARTNERSHIPS

The partnership-building process includes identifying and engaging interests that have a stake in the management of the watershed, establishing an organization, determining a goal and plan for the watershed, as well as implementing the plan and evaluating its impact. Virtually every successful watershed management project to date has demonstrated that local management is essential to stimulate and maintain the interest needed for a successful effort. The partnership's watershed plan provides a living, feasible approach to address the identified problems, which benefit from the support of the people most affected by the recommended approach.

How does a partnership start? Partnerships may begin through individual initiative and grassroots action; through a local, regional, or national nongovernmental organization; or from federal, state, or local government. In general, the reputation, legitimacy, and degree of trust toward the initiator of the partnership appears to affect the potential for success. For example, in March 2000, the Little Conestoga Watershed Alliance held its formative meeting. The driving forces behind this formation were two county residents who decided that a network of persons actively involved in the water-quality management of Little Conestoga was needed. The watershed alliance would provide a forum for landowners, citizens, educators, local government officials, and environmental professionals to work together to formulate a plan to restore the impaired stream. At the initial meeting a number of concerns were identified: high nutrient levels, erosion, siltation, urban runoff, and stream-

Partnership Development and Operation

bank erosion by livestock. The need to rectify the predominant perception of watershed residents that murky waters were ordinary and acceptable was identified as a high-priority action for the alliance to address.[1] A kickoff workshop, particularly with the format of brainstorming in small groups, is considered a key step in focusing the watershed management effort on issues of importance to watershed residents. The kickoff workshop is a good way to start because it can serve as: a media event, to make the public aware of the watershed concerns; and as an educational platform for identifying problems and concerns. It is useful to gain participation from groups other than conservation and environmentalists such as business, education, and industry at the kickoff workshop. This workshop provides the chance to include influential officials in the earliest deliberations.

To be a partner, an entity must be willing to build and maintain relationships. Most likely a small group of stakeholders will come together to address a common concern or issue. The most important outcomes of the initial meeting are a commitment to proceed with a partnership to address the concern or issue and to get others involved. Having a shared responsibility for implementing solutions is an important aspect of being a partner. Key participants from the initial meeting help the partnership development process by recruiting volunteers for the executive and other committees. It is important to build a watershed constituency. A few committed people can start an effort, but they will burn out quickly if others are not joining and sharing responsibility. Any person, agency, or organization, regardless of location, that has an interest in the watershed should be invited to participate in the partnership. It is important to remember partnership development is considered an open-ended process, and that new stakeholders and partners may emerge at any time in the management process. Different people are active at different times and use community-driven goals to motivate, but mutual overall effort remains foremost. Partners are entities that formally commit resources to support the overall watershed management effort, and stakeholders are those who have any interest in the effort and participate in an aspect of it. The partnership's efforts will be more successful if all organizations, agencies, and individuals are involved early, identify shared interests, and work toward a shared vision.

A primary goal of any partnership should be to engage a diverse group of stakeholders who are truly representative of the watershed. The opinions on what is best for the watershed will be diverse because many people share the watershed and will often have different points of view (Box 4.1). Not everyone in the watershed will be interested in participating in the partnership. Some individuals will participate, or show up, because their job involves watershed management. Public expectations and organizational mandates for cooperation motivate other organizations and agencies to become involved in partnerships. The opportunity to improve environmental quality will motivate other individuals. It is important to provide opportunities for environmental groups and others with limited financial resources to be active participants in a consensus-developed watershed management plan. If possible, partnerships should provide financial assistance to these organizations to ensure they are involved in the process. Three distinct groups of individuals who participate in watershed partnerships have been identified:

Those who are affected by, but not interested in, watershed management.
Those who are interested in, but not affected by, watershed management.
Those who are both affected by and interested in watershed management.

The partnership should focus its efforts on finding individuals in the third group. However, most partnerships find stakeholders from all three groups. All three groups need each other to succeed, and the partnership needs to bring them together. Effectively involving the target communities is a difficult, time-consuming approach but one that promises a more successful watershed management effort and a better return on the investment of time and energy than general public participation. These stakeholders can influence the decisions, will be affected by the outcome, will be responsible for final decisions, and in some cases can prevent implementation. The consequences of not being inclusive are extensive. Lawsuits can evolve, plans fall through, people can fail to understand the issue, and all heck can break loose. An active partnership can usually minimize the impact of the latter type of stakeholder actions.

No single group of stakeholders can write a consensus-based watershed management plan alone. Without input from all stakeholders, the developed watershed management plan will be difficult to implement. By seeking consensus, the partnership allows the open expression of all points of view and agreement. All stakeholders must agree that they have been given an equal chance to be heard and that the consensus agreement represents the best solution at that time. Stakeholders can work on a watershed management effort and promote the development of a common plan without demanding common values. In fact, involving stakeholders with different values in working on a common plan to implement those values reinforces the

---

BOX 4.1
The following categories of people, groups, interests, and organizations should be involved in the partnership:

Mass media
Landowners and operators (managers)
Homeowners
Financial institutions
Agribusiness and industries
Farm organizations
Environment and conservation groups
Local elected officials
Federal, state, regional, and local agencies
Chambers of commerce
Educators
Students
Civic and social organizations
Religious leaders
Retired persons
Developers

benefits of considering different perspectives and value systems from the beginning.[2] Ideas stakeholders develop and embrace are not usually personality-dependent and do not collapse with the departure of an individual.

4.3 ORGANIZATION

Meaningful, effective, and positive environmental improvements require a watershed management plan and an organization with the support and resources to carry out the plan.

Institutional arrangements are often overlooked when a watershed management effort is being initiated. Once the stakeholders in the community have been engaged, it is time to organize — to get the process underway. Partnerships organized around only one particular project or problem at a time mean that the groups may be caught up in a series of disconnected projects or making decisions with no end in sight. Organizations formed in response to a crisis or loosely put together tend not to last. To be effective and sustainable, a partnership should be organized and focused. The partnership must engage in meaningful management actions for the long term.

Maintaining the partnership requires work, time, resources, funding, and an understanding that partnerships must have flexible membership. One major issue all partnerships face is the lack of adequate funding support. Partnerships receive financial support in three major categories: organizational development and ongoing support; planning and program development; and program implementation and evaluation. Usually, partnerships face challenges with overall funding stability and flexibility for each funding category, but especially for organizational development and ongoing support. The difficulty with organizations providing funding for development and ongoing support is the lack of products to document progress to justify funding. Funding organizations have to be accountable to their sources of funds and oversight boards. Funding concerns with ongoing support relate to the length of commitment. What happens when funding stops? Does the partnership end? What if outside funding support is required for a partnership to fulfill a vital local need? Finally, a watershed partnership must overtly try to see problems as challenges and opportunities — not barriers to funding. To sustain a credible and long-term effort and be appealing to funding entities, a partnership must have at least the following four characteristics:

A clear goal
A strong management plan with outcomes identified
Leadership representative of the various interests in the watershed
Local support

Holdren[3] and others identified key institutional characteristics of a successful watershed organization. These organizational characteristics include full-time employees, established office space and equipment, established water-quality monitoring and public outreach programs, access to water-quality information, concern for water quality, and interest in citizen participation. Carl Norbeck[4] identified the mechanics for success of local initiatives as: someone to coordinate the startup and seed resources (funding); time to endure the initial phases of team building; clear expectations about the group being a cooperative endeavor; willingness of agencies to work with bottom-up efforts; and some early successes that validate the group's efforts. The watershed management process requires a lot of coordination. It is important to have a focal point for the watershed effort and ensure someone is paying attention to moving committee activities along. These partnership organizational aspects and activities should not be confused with the processes for actual watershed planning, but their direction setting will lead partners to decide what they need in their watershed plan.

In a workshop on watershed initiatives, Born and Genskow[5] highlighted the arrangement favored by workshop participants is for a paid coordinator working directly for the partnership to lead the process. This provides the partnership with a point of focus. For many watershed management efforts, agency staff will be in positions of coordinating and leading the partnership effort. In other circumstances, agency staff will simply hold one of many seats at the table on various committees and teams. The role of the watershed coordinator is to coordinate and monitor all project activities. This role requires someone who can integrate individual activities into a total project. The coordinator needs to have a clear understanding of specific project goals, possess the various strategies to reach those goals, and know how these strategies will come together to form an effective project. Primary qualifications for the watershed coordinator include excellent organizational and interpersonal skills and the ability to work closely with individuals and groups. Strong communication skills, including writing and public speaking, are also necessary to build public support within the watershed. In addition, the watershed coordinator needs to have sufficient technical background to understand the water-quality problem and watershed management strategy and be able to communicate this information to the general public as well as targeted audiences. Finally, in order to facilitate coordination, the coordinator has to be familiar with complementary goals and opportunities available through other natural resource and water-quality programs.[6]

4.3.1 STRUCTURE

No one-size-fits-all organizational model would work for all watershed partnerships. The form and structure of a watershed partnership can range from an informal organization to a more complex, formally organized structure (continuity in watershed management is critical, and a formal organization helps ensure that continuity). The right type of organization depends on a number of factors, including the geographic scope of the effort, whether an existing organization can fulfill the role, and the resources available. It is important to make sure that "form follows function." Both the structure of the watershed management organizations and the type of activities supported need to be matched to the scale of the watershed and its issues and concerns. The structure the partnership selects must be tailored to the unique social circumstances of the watershed. Some division of labor and delegation of responsibility are needed within the partnership to take advantage of available resources and expertise. However, a minimum organizational structure is recommended for a community-based, voluntary watershed management effort. This con-

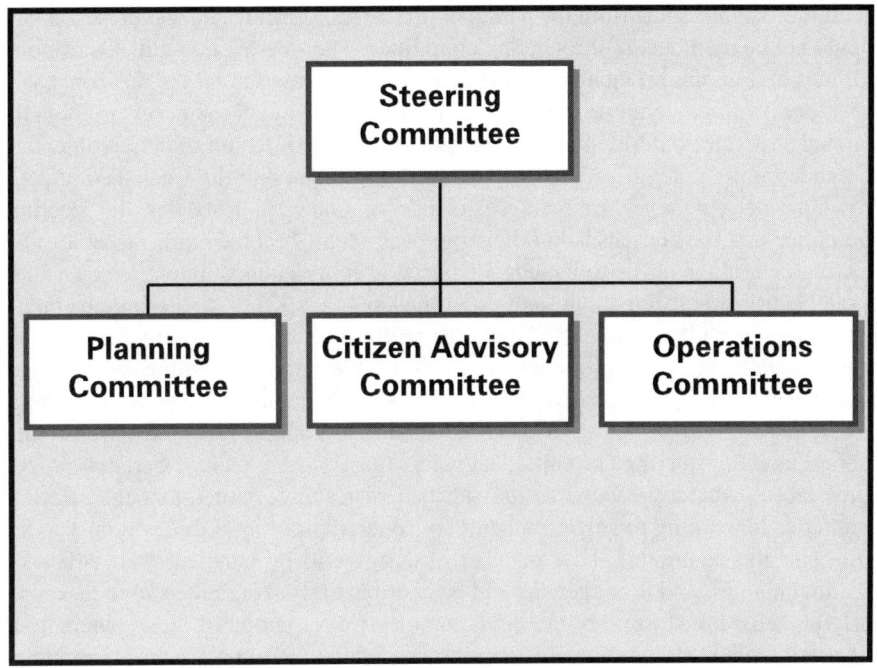

FIGURE 4.1 Partnership organization structure. (Courtesy of CTIC, West Lafayette, IN.)

sists of the following committees: steering, planning, citizen advisory, and operations (Figure 4.1). Other possible committee or subcommittee structures would cover activities such as fund raising/finance, evaluation, outreach, communications, implementation, monitoring, and community relations. Too many committees or committees that are too large are often ineffective for managing watershed plan development and implementation. Each project must find the right balance or mix for its watershed management efforts. The public should have the opportunity to be involved in any and all committees or subcommittees formed. Not all stakeholders need to participate in all aspects of watershed initiatives. Balance is needed in determining the composition of the various committees, as well as the total number of committees.

The first committee that needs to be formed is the steering committee, also known as the management or executive committee. A committee is a collection of individuals representing a specific group of individuals or interest in a matter of mutual concern or interest. Successful partnerships do not just happen; they depend on the leaders who emerge from the stakeholders — leaders are needed on the steering committee. Leadership is a critical factor in making the watershed approach work. Leadership is traditionally viewed as a role for an individual, but the role can come from a group or entity, such as a local board. One of the roles of the steering committee is to provide leadership for the overall watershed management effort. The steering committee should be a relatively small group, less than a dozen people who are interested in the watershed, who are willing to volunteer, and who represent primary interests in the watershed and provide diversity. Leadership of the steering committee should come from the public or private sector, and agency representatives should not be members of the steering committee. The steering committee's responsibilities at a minimum include setting project direction based on the mission statement, performing overall project management, and setting up the necessary organizational structure to develop and implement the plan. Exclusion of certain interests can undermine the legitimacy of the partnership or even halt the watershed effort.

Conversely, a large group might include so many interests that the steering committee and the consensus-building approach might become unmanageable. The steering committee needs to include a balance of representative interests consisting of public, private-citizen public-interest groups, public officials, and economic interests guiding the process to establishing the partnership's mission statement and the project's goals. Steering committee members must keep their constituency informed about the watershed effort, actively interacting with their constituency and accurately representing group positions to the watershed partnership.[5] Operating procedures address how the steering committee members function as a team. Operating procedures are not directly related to the substantive resource issues that the steering committee is working to solve. Rather, they concern procedural things such as who will chair the committee, how member absences will be handled, who will take meeting minutes, and how agendas will be distributed. Making procedural decisions early on helps transform the steering committee from a group of individuals into a cohesive, organized committee. Operating procedures help the steering committee stay focused throughout the watershed management process on the problems and solutions. One major issue that might have to be addressed in the development of operating procedures is transboundary coordination. Given the transboundary nature of watersheds, which do not correspond to government jurisdictions, state agencies must be able to provide the requisite coordination with federal agencies and state and tribal governments.

Steering committee membership, roles, and responsibilities vary depending on water-quality issues and interests of the individuals. Effective steering committees generally coordinate committees' activities and keep the partnership moving forward. The steering committee handles or delegates administrative details to various committees or partners. An important shared role of the steering committee is identifying program resources available to address watershed management needs. The challenge before the steering committee is to reach consensus on programmatic elements that will address the unique needs across the watershed in a logical and cost-effective manner.

Once the steering committee is in place, the next step is to establish the other committees and designate a lead agency or organization. The lead agency's role is to ensure that the watershed planning and implementation process continues to move forward. The Great Swamp Watershed Project is managed by the Ten Towns Great Swamp Watershed Management Committee under an interlocal agreement from the ten municipalities in the watershed.[7] Committee participants usually belong to one of three groups. The first, and most predominant, consists of government agency employees who participate because it is their job. They are not recommended for membership because it would limit their ability to pursue their own agenda. A second group consists of organizations whose constituents have an interest in the issues being addressed. The third group consists of stakeholders, but do they fully represent the public?

The partnership's mission statement must answer the following: who the members are; why it exists; and what it stands for. The latter two components provide direction for the goal-setting process and focus for the planning committee. The process of establishing and adopting a mission statement ensures common understanding is reached early in the project development. The best mission statements are graphic in their descriptions and relate to human experience.

The mission statement articulates the focus and direction of the partnership's efforts. A clear mission statement helps stakeholders understand, relate to, and support watershed protection and restoration efforts. In addition, the mission statement informs the general public, elected officials, business, the press, and community leaders in the watershed about the partnership and the overall scope of their efforts. The mission statement should be used in all the promotional material, on correspondence, in the watershed management plan, and in funding applications. The following hypothetical examples of a partnership mission statement illustrate the range of issues covered and the level of detail.

> The Lake Ethan Partnership has a vision for the watershed that will:
> Maintain, protect, and enhance a balanced watershed fishery
> Preserve and enhance an ecosystem that supports a diverse and balanced wildlife population
> Include water quality suitable for a full range of recreational opportunities
> The purpose of the Emily Creek Partnership is to develop, enhance, and protect the ecological and socioeconomic values of the watershed's natural resources while continuing private ownership.

After the partnership establishes its overall goals, an acceptable planning process must be set up. In order to develop a plan that addresses all resource concerns and integrates ecological, economic, and social factors, multiple stakeholders interested in developing a watershed management plan need to be identified. While the steering committee manages the overall process, the planning committee develops the watershed management plan, working closely with the technical advisory committee (TAC). Advisory committees make recommendations to another committee or higher authority. The recommendations are not binding. Make sure all committee, team, and workgroup members fully understand the purpose and goals of the management effort. The TAC completes the watershed assessment and provides the technical basis for the watershed management plan. The planning committee determines the objectives after identifying the resource concerns in the watershed. The objectives become the goals the planning committee intends on accomplishing. The TAC uses the objectives to understand what the planning committee wants to accomplish, and they identify possible solutions to the identified problems that can meet the partnership's objectives.

An effective way for a partnership's steering committee to rapidly learn about the watershed's issues and condition is to conduct a rapid resource appraisal (RRA). The RRA helps the committee identify the resource concerns, define objectives for the watershed, and learn about the ecological, social, economic, and political aspects of the watershed. An RRA usually takes a full day and consists of the following activities: educational presentations by technical specialists; watershed tour with tour stops; discussion sessions with stakeholders; and meetings with government officials and agencies. By engaging in discussions and learning together in the watershed, the committee starts to develop a common vision and an identity for the partnership. Visual inventories can provide a great opportunity to involve the steering committee in the project and enhance its familiarization during the assessment phase with the watershed. The RRA should be the first step in completing visual inventories of the watershed.

The planning committee usually consists of members from the steering committee and members from other interested organizations and agencies. There should be no restriction on planning committee membership. Planning committee members should be able to: collectively represent a special-interest group as well as their individual interests; serve as decision makers in the watershed; together represent all the social, economic, and cultural communities in the watershed; and represent all the different views, opinions, and interests in the watershed. Using the mission statement, the planning committee defines its purpose and the purpose for the watershed management planning effort and implements the partnership's planning process. The planning committee will use the information regarding the water resource condition to evaluate proposed solutions to select the alternatives that will provide the best framework for the implementation plan. Working closely with the TAC, the planning committee sets the boundaries on the assessment phase. In addition to the overall planning committee, some watershed partnerships form subcommittees or workgroups to develop specific parts of the watershed plan. Subcommittees and workgroups are formed along administrative lines or along objective lines. The planning committee develops a work plan to guide the development of the watershed management. It is important to remember the development of watershed management goals is considered an open-ended process initiated during the assessment phase. New goals may emerge at any time during the planning and implementation process. The watershed management work plan serves two main purposes: as a communication tool, and as a tool to project schedules and work loads for developing the management plan. The work plan provides a basis for operations and progress evaluations. The logistics for completing the work ahead have to be estimated by the planning committee and included in the work plan. The planning committee needs to develop a solid, workable plan and review it annually.

The TAC is a team of professionals and interested stakeholders, usually between 20 to 25 individuals, who assess and evaluate available information and data. The TAC also makes recommendations concerning the need for additional data and information and then, during the planning phase, suggests management strategies and approaches that should meet the goals and objectives developed by the steering and planning committees. The TAC uses the information about what is known and what needs to be known to set priorities for the inventory and assessment work. The TAC membership usually consists of experts and representatives from nonprofit organizations, local organizations, state government, universities (public and private), and the federal government. The interdisciplinary TAC should be organized to draw upon the knowledge and skills of different agencies, organizations, and individuals. The TAC needs to use an interdisciplinary approach that has all the key disciplines (biology, hydrology, etc.) equally involved in the assessment/problem identification and formulation of alternative strategies process. Therefore, one of the first steps in the watershed assessment and identification phase is the determination of the proper disciplines for the effort.

The TAC has an initial meeting where members learn about the planning process, the perceived watershed issues and concerns, and their role in the overall effort. The TAC should select a "chair" at the initial meeting. The role of the chair is to coordinate TAC activities, work with the planning committee, and help facilitate communication among TAC members. TAC members usually complete their assignments individually or in small groups or teams and periodically report results to the entire TAC. The TAC teams can provide continuity as well as important information and insight from varied disciplines, experiences, and backgrounds. The individual members determine if they complete their work individually or use a group process. In formal organizations it may be necessary for the TAC to write a "plan of operation" to guide its work. This plan of work documents the watershed issues and resource concerns each TAC member will be involved with, the inventory and evaluation activities to be carried out, and schedule for completion. In addition, TAC members are responsible for overseeing or conducting any additional resource inventory and data collection efforts needed. TAC members are responsible for determining what is technically feasible to accomplish.

Some watershed partnerships create citizen advisory committees (CAC) to provide advice on various aspects of the partnership operations. CACs are most appropriate when the input, influence, and involvement of a large portion of the watershed residents are needed. In other situations, there will be times that for a particular issue a partnership might create a CAC to obtain focused input. When forming an advisory committee there needs to be a commitment to take their advice seriously. Develop a set of core principles to use when dealing with citizen advisory groups to ensure that personalities and personal feelings do not damage the group's potential. Consider creating a CAC when developing potential solutions to controversial problems, seeking public reaction, wanting help to monitor program implementation, and demonstrating accountability, openness, and responsiveness. When creating a CAC, the partnership needs to have the CAC tasks outlined with responsibilities established. This allows potential members to understand that their role is advisory, not decision making. Realistic estimates of the time and resource requirements of an advisory committee for both the steering committee and potential advisory committee members must be given. The steering committee has to ensure enough support is available to make the CAC successful. CACs are increasingly recognized as an important aspect in successful watershed management projects. Many factors determine the success or failure of a CAC — committee size, how committee members are chosen, agency staff and steering committee members' attitude toward CAC involvement. It is important that a record of all advisory meetings is developed and available to the public. One major factor in ascertaining the success of a CAC involves how the members relate to and interact with each other in committee meetings. Committee members need to respect each other's opinions.

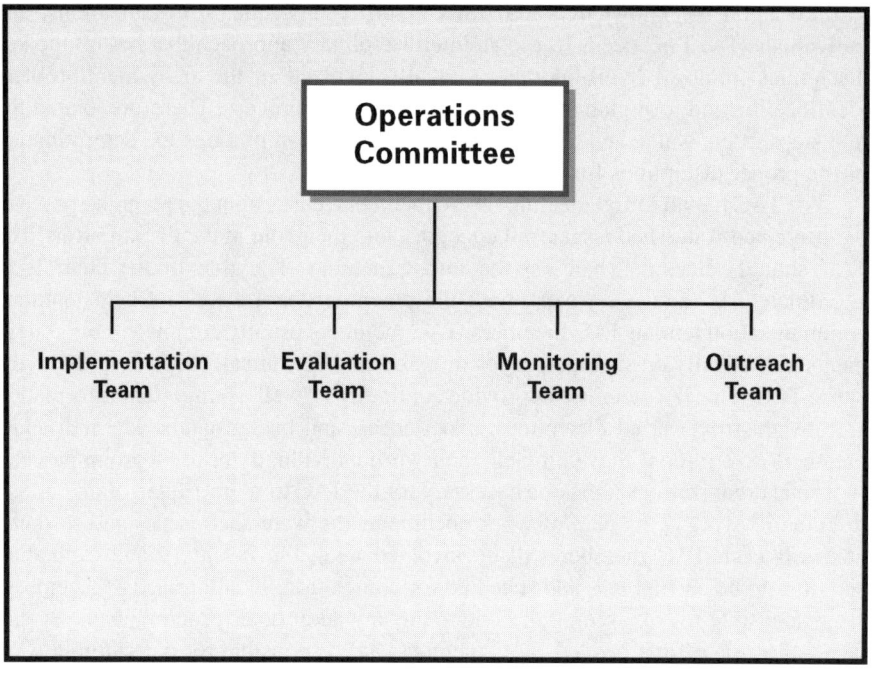

FIGURE 4.2 Operations committee and its teams. (Courtesy of CTIC, West Lafayette, IN.)

The CAC is a means of getting input to and assistance for the watershed management process. Partnerships should consider public hearings for clarifying issues; gathering information, positions, and opinions; and providing an opportunity for people to vent. The steering committee often faces stakeholders whose ideas range from skepticism regarding big government and a hands-off attitude to environmental extremist and government control. Citizen advisory committees instead of the steering committee can be the focal point for this type of stakeholder participation. Neighbors talking to neighbors is usually more effective than steering committees seeking support. Stakeholder participation in government-sponsored programs can result in a more informed public, improved government, and the best possible decisions for proposed actions or plans.

The operations committee is responsible for implementation, evaluation, outreach, and monitoring. Figure 4.2 highlights the administrative operations of the implementation aspects of the partnership. Depending on the scope and complexity of the implementation activities a number of operations teams or workgroups may need to be created to provide a focus for the various activities. At a minimum, workgroups should be formed for each function. The majority of successful watershed management efforts include a formal organization to develop and implement the outreach component. The planning committee determines the primary message, and the outreach team decides on target groups and the best strategies for reaching them. An important part of the outreach component is convincing stakeholders that they are stakeholders. It is important that the outreach effort is closely coordinated with the TAC and planning committee activities.

When completed studies, reports, and other information become available to the public, the partnership should consider creating a document repository. The operations committee would be responsible for the repository. This would be for assessment reports, planning documents, and implementation and evaluation reports. When studies and reports are summarized in a media release, fact sheet, or meeting notes, the public can be referred to the repository. A document repository should be a public location where interested people can gain access to program-related documents. When considering repository locations, consider not only the location, but also hours of operation, access for disabled people, proximity to public transportation, and availability of photocopying.

4.3.2 Partnership Operations

The partnership needs to adopt an operating process and ground rules in order to effectively and efficiently manage meetings and the decision-making process. While our government is based on an adversarial system, where proposals are made in public meetings and different groups of people advocate for or against the proposal, in most situations it is not a good example for public participation for watershed management plan development. In some instance's "Robert's Rules of Order" traditionally run meetings can be an impediment to open frank discussion and reaching agreements. Many democratic methods that foster a win–lose environment like voting may actually hinder reaching consensus. The win–lose environment results in a minority that may never truly buy into the proposed solution. A true stakeholder participatory process has the stakeholders not only advocating positions and debating issues, but also listening to a diversity of viewpoints, and developing compromises and solutions. It is not about tradeoffs or compromise. It is about respect and genuine support for recommendations and solutions to which everyone agrees. Seeking a consensus is best when all stakeholders need to support and implement the solutions. Consensus building is procedurally more burdensome to manage and usually more time-consuming, but it does provide a foundation for action and participation.

The challenge before the stakeholders is to reach consensus on management approaches that will address the unique needs across a watershed's various land uses in a logical and cost-effective manner. More watershed management efforts fail due to social and political reasons than the lack of a scientific solution. Consensus building helps avoid this problem. Consensus decision making is the most effective decision-making method that a partnership can utilize; however, it is not easy and takes commitment and time. What is consensus? It is a process that results in a decision that everyone can live with and everyone agrees to support and work toward.

Effective consensus decision making shares the following characteristics:

1. Total participation. All major interests are identified and included in the partnership. The partnership is actively involved in maintaining a broad range of stakeholders and citizens in the planning and implementation of the watershed management efforts.

2. All partners are responsible. Everyone assists in planning and implementation activities and offers suggestions to make them more practical and effective.
3. Partners educate each other. Partners spend time discussing the issues including the history from their perspective; their perceptions and concerns; and possible solutions. Each partner has the opportunity and responsibility for meaningful contributions.
4. People are kept informed. Participants keep their own groups and the partnerships informed. Partners document, publicize, and celebrate the successes through an ongoing recognition program and communication program. (See the social capacity building chapter for more information on developing and implementing a communication program.)
5. Multiple options are examined. Partners seek a range of options to satisfy their respective concerns and avoid promoting single positions.
6. Decisions are made by mutual agreement. Partners do not vote but modify options or seek alternatives until everyone agrees that the best decision has been reached.
7. Partners are responsible for implementation. The group members identify ways to implement solutions.

How do you reach consensus?

1. Establish the majority opinion
2. Examine all minority views, checking for understanding
3. Explore ideas to achieve a decision or position acceptable to all
4. Restate a new majority opinion
5. Establish the new majority opinion that becomes a consensus

Successful partnerships take time and commitment to develop. There are four main stages in the partnership development process. Each stage takes time and involves specific actions and feelings. Additionally, there are three aspects of a successful partnership: (1) there is buy-in by partners; (2) the partnership creates a framework with partners' ongoing needs; and (3) by making adjustments partners can usually make a big difference. Knowing these stages will make it easier for the partnership to move forward and reach success at the end.

1. Forming. Members cautiously explore each other. Feelings during this stage include skepticism and anxiety as well as optimism and excitement. Activities completed during this stage are establishing organization, goals, objectives, and work plans; determining what data and information needs to be collected for the project; and discussing issues, concerns, and concepts.
2. Storming. Partners become impatient and begin arguing. Feelings include resistance to change and negative attitudes about the success of the partnership and project. Partners exhibit the following behavior: argue about minor issues and concerns; become defensive; and wish to revisit existing agreements. Activities occurring this stage include discussing alternatives

Partnership Development and Operation   51

and impacts and developing unrealistic goals and proposed activities. Partners become competitive, and tension and jealousy increase.

3. Normalizing. Conflicts are reduced and partners become cooperative. Partners accept their roles as well as partnership norms. Feelings include acceptance of team membership. Partners exhibit the following behavior: achieve harmony by avoiding conflict; exhibit more friendliness and willingness to share problems and opportunities; and show a sense of team cohesion and common goals. Activities completed during this stage include plan development and implementation begins. Commitments from partners occur. Cooperation requires some recognition on part of each organization that the common interest is best achieved by some cooperative effort. Cooperation occurs when organizations share information and interact regarding specific tasks to benefit their clients or to achieve common goals.

4. Performing. The partnership has become an effective and close-knit group. Feelings include new insights about the partnership and each member's role as well as satisfaction with the partnership's progress. Partners exhibit the following behavior: constructive change; ability to work through issues and problems; and close attachment to the partnership. Activities completed during this stage include implementing and evaluating the plan and identifying new opportunities. As partners sustain communication and are able to understand more about each other, a number of critical choices present themselves. Similarities and differences emerge such as differing mandates, accountability, partnership flexibility, autonomy, leadership, partnership cultures, and partnership alliances. The new knowledge can create the potential for conflict (storming) as well as movement toward cooperation (normalizing and performing). One voice must not be dominant. All voices need to be heard, all positions identified, and all points of agreement acknowledged during the consensus-building process. Figure 4.3 shows the continuum of partnership relationships from the initial meeting to successful implementation. Cooperation builds from the points of agreement by changing policies and strategies. Cooperation requires some recognition by each partner that there is a common interest best met by some cooperative effort. Cooperation occurs when partners share information and interact around specific tasks to benefit the partnership or to achieve common goals. The primary signs of stability in a cooperative partnership include: institutionalizing information exchange; agreeing how to handle conflicts among partners; and sharing an identity as expressed in a mission statement or some other statement of their vision.

Coordination is attained when partners agree on strategies that modify existing services or activities to achieve more efficient and effective service consistent with the missions of their respective organizations. Coordination as represented on the continuum is broader than cooperation. Partners interface resources and they may need to change their own policies and practices in order to accomplish a task or deliver a service. They make long-term shared commitments based on the watershed

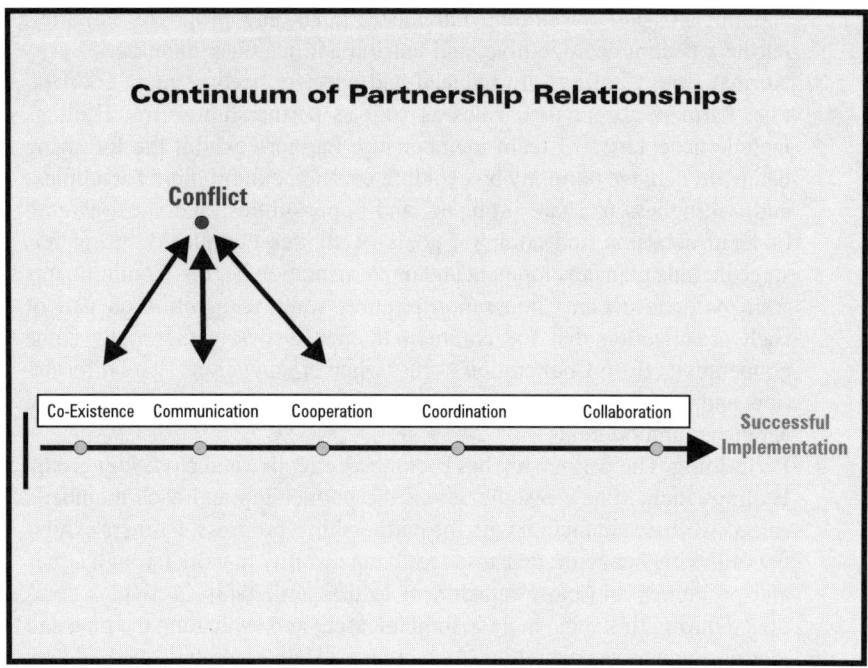

FIGURE 4.3 Continuum of partnership relationships. (Courtesy of CTIC, West Lafayette, IN.)

management plan. However, they do not fundamentally change the way they do business or make decisions. In collaboration the partners change the way they do business by placing the project's needs over the bureaucratic culture.

Most people agree with the notion of partnership, at least in principle, but that does not mean they will fully participate or that the partnership will be successful. The Conservation Technology Information Center (CTIC)[8,9] identifies a variety of reasons a partnership may be unsuccessful.

1. Lack of commitment. Partners have other commitments and limited resources, so it is important to ensure the partnership is working toward a worthwhile goal. For example, in the late 1990s, the USEPA's Region 5 Water Division embarked on a watershed project in Northeast Illinois to protect a trout stream; unfortunately, the effort failed. Changing priorities, lack of local stakeholders, and inability of management to comprehend the scope and need for long-term commitment resulted in the USEPA moving on and the project being abandoned by the other federal and state agencies who were supporting the USEPA's effort.
2. Past failures. Acknowledge past failures and use them when setting up the partnership. The history of unresolved conflict among members and unwillingness to work at resolving the conflict must not be viewed as a barrier. The most important thing is to learn from failure and not let it be

an omen of the future. For example, some managers agreed to everything but do not commit the resources needed to fulfill the commitment.
3. Concern about lost independence. Partners need to define their reality for working in the partnership. Partners need to clearly identify their commitments and the limitations of their involvement. All partners need to know they are working toward a worthwhile goal. Partners also need to know what is expected of them.
4. Personality conflicts. Select partners based on existing and potential skills, not personalities.
5. Lack of credit for contributions. Some partners do not receive credit for their contributions. Partners need to get credit for their contributions and keeping their own groups informed. The watershed management plan's communications plan needs to clearly identify roles and responsibilities for reporting achievements and how it will be done. For example, a number of early Darby Creek organization publications failed to list the USEPA as a partner, because it was providing its funding and support indirectly through other partnership organizations. This failure became a distraction in the development of future efforts because a focus of the work plan became ensuring the USEPA got recognition for its contribution.
6. Power struggles. Some partners have a disproportionate amount of power. Partners need to guard against using their authority and expertise to hinder the consensus-building process by exercising too much influence over the partnership. Some partners, in particular regulatory agencies such as the USEPA and USACOE, may want to use the partnership to pursue their own agenda rather than the goals of the partnership. This does not mean partnerships cannot be used to facilitate a regulatory agenda or effort. For example, the Superior (Wisconsin) Special Area Management Plan (SAMP)[10] was developed to protect and preserve high-quality wetlands in balance with sustainable development. Working with USACOE–St. Paul District and USEPA–Region 5, the partnership developed a SAMP that streamlined the regulatory process to expedite permit decisions. The SAMP development process took nearly seven years and involved a major commitment of time on the part of the city, the regulatory agencies, and other partners. Since the SAMP has been in place, the economic development benefit has been considerable, and the SAMP is now used as a marketing tool by the City of Superior. Other partners may look toward the partnership as a means to avoid or get out of regulatory commitments and requirements, which is also inappropriate. Partnerships need to target and acknowledge the differences in the vested interest of each partner and work with them.
7. Turf battles. Partners with historic feuds over areas of jurisdiction and program responsibilities need to work in a cooperation- and coordination-building mode to ensure they do not continue the feud in the current partnership.
8. Partners that do not agree on realistic roles, responsibilities, goals, or timeframe. This usually results in a lack of follow-through with commit-

ments. Decisions regarding commitments must be made by mutual agreement and not as individual expectations or by other agencies and organizations. Collaboration in a partnership requires a fundamental change in the way partner organizations and agencies function. Some decision making and planning are shared, and partners commit resources to be overseen by the steering committee that represents the partnership.
9. Differences in cultural and personal values. Partners need to think as an entity or group and not as individuals. Informal, social interaction can provide the foundation that holds the partnership together.
10. Communication is insufficient. Existence of communication in a partnership does not imply any other type of interaction. For partnerships to be capable of doing any work together, the partners must move from the communication stage to the cooperation stage.
11. Key interest not included or refusal to participate in the partnership.
12. Problems are not clearly defined or not felt to be critical; therefore, the focus lacks a clear purpose.

As part of the partnership-forming process, conflict can be expected between stakeholders. Figure 4.3 highlights when conflicts can be most often expected. Cooperative partnerships are vulnerable to conflict and will be unable to do productive work until they develop an identity as a partnership that is separate from and inclusive of their independent organizational identities. It needs to become "we," and the partners need to complete some simple tasks successfully and gain some experience working together.

A minimum level of trust must be established so that all partners will respect minimum operating rules and develop mechanisms for recognizing and resolving conflicts. In the watershed management process conflicts often arise. This occurs when partners' policies conflict, when social values conflict with environmental needs, or when planning recommendations conflict with existing polices and approaches. The five basic conflict management strategies are avoidance, accommodation, competition, compromise, and collaboration. The partnership, committee, or team must analyze the conflict and select the most appropriate strategy to address it. There are four basic principles to help partnerships avoid destructive conflicts and promote constructive interaction. Where conflicts cannot be avoided, these principles minimize the fall out from it.

The four principles are:

1. Preserve everyone's dignity.
2. Listen with concern.
3. Be flexible while maintaining your independence.
4. Do not expect people's personalities to change.

The discussion that follows provides an overview of conflict resolution and management. If the partnership is in an extremely hostile situation, it may require skill beyond the partnership's capabilities to resolve. It should not hesitate to get help from someone who specializes in conflict mediation. Resolving conflicts is not

Partnership Development and Operation

a mechanical process in which applying steps one through six, listed below, guarantees resolution. It is important to note that not all conflict is resolvable. Conflicts based on value systems or personalities are the most common unresolvable conflicts. Conflict is a normal part of human interaction. Anytime a shared resource is at stake, you can expect differing views regarding the importance, use, value, or investment in that resource. In many instances organizations and agencies will have formal positions regarding a resource and its use. Rather than always working to avoid conflicts, it may be more efficient and effective to manage conflict. Conflicts cannot be ignored. A number of negative aspects are associated with conflict within a partnership: partners can become less motivated and inactive; intergroup tension can increase and divert attention from the objectives; antagonism can increase; partners may seek safer subjects and not address difficult or controversial issues; and stress and frustration can occur. Successfully resolving conflicts does have positive aspects: it increases partnership cohesion; partners learn more about the individuals involved; and the partners grow with the conflict experience. There are various management strategies for addressing conflicts: ventilation, active listening, reduction, brainstorming, and compromising. CTIC[9] in *Managing Conflict, a Guide for Watershed Partnership* covers understanding and managing conflicts, as well as negotiation skills in the context of watershed partnership dynamics.

The partners' differing views and positions concerning the waterbody need to be identified and worked with; address them openly and respectively. Focusing on interest rather than position makes it possible for the partners to come up with better agreements. Remember that people's feelings are just as real to them as facts are. It is important to note some of these conflicts may be related to historic or personal difficulties between individuals and groups. Conflict is always a possibility in a partnership setting (Figure 4.3). Conflict is not always apparent or overt. At the early development stage of a partnership, conflict can destroy fragile beginning relationships. Initial partnership efforts to work together are often vulnerable to conflict, and relationships may go in and out of conflict in the early stages. Two types of conflict have been identified in watershed partnerships: destructive and constructive. Destructive conflicts are generally unresolved. They may be real or perceived, provide the foundation for defensiveness, reduce communication, and can even cause the termination of some of the working relationships within the partnership. Constructive conflict can result in relationships becoming stronger; parties become more effective at working out their differences; parties begin to work in an atmosphere of trust; the parties having the conflict are satisfied with the outcome; and all parties have improved ability to resolve future conflict. A six-step process to manage and minimize a conflict follows:

1. Identify and define the conflict. Determine the conflict's content and history. Pay particular attention to personalities and positions.
2. Brainstorm. Solicit discussion, information, etc. from all partners.
3. Evaluate the information. Make sure everyone understands the facts and feelings behind the conflict.
4. Choose a solution. Identify common goals and values related to the issue or concern. Document the areas of agreement. Offer a compromise for

the unresolved issues. Review any proposed agreements, and make sure everyone understands them. If a compromise cannot be reached, table the issue and move to the next issue. This rule for consensus is useful: if on a particular issue, stakeholders still disagree after extensive discussion, the stakeholders must be able to agree that everyone has been given a chance to be heard and the consensus agreement represents the best solution at the time.
5. Implement the solution. Solidify agreements and begin to implement the agreement.
6. Follow up. Check to make sure everyone is abiding by the agreement.

It is important just to spend time and resources maintaining the partnerships. A number of activities can be done to maintain and strengthen the partnership. They include:

- Start with small projects that will provide early successes.
- Keep the partners focused on a common goal and task before the group.
- Respect participants' time.
- Document and celebrate progress.
- Build ownership for the partnership at all levels of the partner organization.
- Maintain a stable structure, with accountability to partners.
- Identify specific benefits to stakeholders and participants.
- Never blame. Always allow participants a means of saving face.

The primary signs of stability in cooperative partnerships include institutionalization of information exchange and agreement about how to handle conflicts among partners.

4.4 INSTITUTIONAL ASPECTS

The partnership has to determine how formal it needs to be in order to function. For example, if the partnership is going to be responsible for directly receiving funds, manage consultants and contractors, and obtain permits, it should create an organization to handle this. If a partnership has designated a lead agency to handle the administrative and oversight functions, it can be less formal. If designating a new or existing organization significant, new responsibilities, the partnership needs to make sure it has the capacity and resources to do the work.

Another issue that needs to be addressed during this process is the need for a formal signed agreement between the partnership and government agencies. Formal agreements provide partnerships a bridge to agencies during reorganizations and staffing changes. In spite of the requisite staff resources that agencies can bring to a watershed management effort, they cannot dominate and make a mockery of the partnership concept. The most common misstep agencies make today in soliciting participation is presenting a partnership a watershed management plan prepared by professionals without their input. This violates the basic premise of public participation and seldom results in community ownership of the plan. Formal agreements attempt to avoid disappointment and misunderstanding. Formal agreements should list the specific actions to be taken by the agencies, include deadlines for each action, include a conflict resolution process should misunderstanding occur, and include a definitions section for terms that may mean different things to different organizations and agencies.

It is important to keep in mind that individuals attending the partnership meeting represent and make the commitment for their respective agencies. In many agencies, such as the USEPA, midlevel managers and staff are unable to make long-term commitments to efforts due to the lack of authority or technical expertise concerning what needs to be done. Due to the long time frames for watershed management efforts, it is important to gain the support of the resource decision makers within agencies, for example, the regional administrators for the USEPA, and the state conservationist for the USDA–Natural Resources Conservation Service.

Watershed partnerships may have limited ability to influence upper-level agency administrators, but their ability to involve agency champions to pursue and secure agency support plays an important role in the partnership's ultimate level of accomplishment. In addition to the aspects mentioned earlier, formal agreements with agencies also help address the issue of conflicting messages from local, state, tribal, or federal agencies participating in a watershed management effort which often then results in low rates of participation. Each agency must clearly define its role and how the agency will interact with other partners to avoid confusion, duplication of efforts, or competition. Agency administrators need to express support for the effort and emphasize the need for partnership communication and cooperation.

4.5 POLITICAL INVOLVEMENT

The first part of developing political support for the watershed management effort is to gain an understanding of the community interest in the water resources. Community benefits from watershed management often include improved implementation of traditional delegated water programs, improved certainty and predictability for residents, and improved community ability to allocate lands to their most appropriate or suitable uses. The goal is to make the watershed management efforts visible to the general public and to put local political support in a situation where they will gain in public image with their support. A number of variables such as money and threats to the environment enable implementation. But the greatest variable is the power of the concerned citizen that can tip the scale most often to support implementation.

"Tip" O'Neil, former speaker of the House, once said, "All politics are local," and to that can be added "and all watershed management efforts are passionately political." The term "politics" itself gives an indication of the problem, "poly" meaning many and "tic" referring to the blood-sucking burrowing insect. The biggest challenge most partnerships face is bringing local officials with diverse ideas together to develop support for watershed management efforts that transcend their political boundaries.

Local authorities are making decisions every day regarding land use, zoning, and new developments that affect the quality of the watershed. Local government officials' involvement and support have been identified as a key to the success of watershed management planning and implementation. Obtaining the enthusiastic support of one or more influential politicians is crucial to most land-use planning and implementation efforts. This can make everything else come much more easily. In order to develop your strategy on how to gain elected officials' involvement and support, it is important you have an understanding of their possible reasons for participating. Coastlines[11] summarizes portions of a study on local government officials' participation in watershed management planning. The study conducted by the Social and Environmental Research Institute located in Leverett, MA, focused on local government officials from three national estuary projects. The first phase of the study covered their reasons for deciding whether or not to get involved with their respective estuary projects. The second phase of the study ranked the items in order of importance for their involvement. The third phase analyzed the data from the first two phases to find patterns. An analysis (inverted factor analysis) of the ranked order of importance identified five patterns of beliefs that were then translated into general perspectives. These general perspectives, presented below, are not exclusive in nature and need to be viewed in the context of the local political and social conditions.

- Perspective one. Influencing outcomes through engaged, effective interaction with others. Individuals with perspective are concerned about the capability of the project to achieve results and the working relationships with project participants.
- Perspective two. Finding the time in a busy schedule. There is a need to show the relevance to the community. Individuals with perspective participate in watershed efforts to provide direct benefits to the community and address problems facing the community.
- Perspective three. Solving problems and serving community. This perspective reflects a problem-driven approach to deciding whether or not to participate, which is very much locally based. For example, protecting the community's drinking-water supply from encroaching development would advocate for involvement.
- Perspective four. Moral compulsion, power, and effectiveness. This perspective is characteristic of an individual who believes that his or her involvement will have a positive impact. To these morally driven individuals, the results matter, not the quality of the process.
- Perspective five. Matching skills, experiences, and position with the project. This point of view reflects whether there is a match between the individual's ethics, experience, and personal interest and the watershed management efforts.

These various perspectives indicate that there is no one way to approach local government officials. To maximize local government officials' participation in a watershed management effort, it must be presented in a variety of ways that highlight different values, needs, and beliefs for different officials. Since time constraints are

Partnership Development and Operation

a limiting factor for local government officials' involvement, a variety of levels of participation needs to be presented when approaching them for their support. Positive approaches for involving the politicians include:

Use of advisory committees
One-on-one contact
Workshops
Use of public comments, procedures, and public hearings

In *Restoring Streams in Cities* Ann Riley[12] shares a valuable lesson she learned from community organizing experts concerning arranging meetings with local public officials to gain their support. When meeting with pubic officials, no fewer than five citizens should participate in the meeting to discuss the effort. The reasons for the critical mass are to demonstrate diverse support for your efforts and present the plan as a group, which creates a different dynamic with local public officials than if just one or two individuals approach the officials. Creating this dynamic should result in your efforts seen as higher priority for local officials' involvement and support. In demonstrating diverse support include a combination of civic, social, and business organizations as well as traditional environmental groups. Follow up the meeting with invitations to participate in high-profile events.

It is always beneficial to obtain congressional support and involvement in your watershed management effort. In addition to assistance in obtaining financial support from various agencies or directly through pork barrel mechanisms, congressional staff can be very helpful in facilitating your interactions with federal agencies. Congressional involvement will not result in watershed management efforts' circumventing existing laws and regulations. However, congressional involvement and interest will increase the level of attention and involvement federal agencies provide to the effort.

4.6 CONCLUSION

The questions a watershed partnership should be able answer are:

1. Who are we?
2. What are we about (mission statement)?
3. What do we do?
4. How do we function?

For more information regarding the stages of partnerships and conflicts, see:

CTIC, *Building Local Partnership, a Guide for Watershed Partnerships,* Know Your Watershed Kit Guides, CTIC, West Lafayette, IN, 1995.
CTIC, *Managing Conflict, a Guide for Watershed Partnerships,* Know Your Watershed Kit Guides, CTIC, West Lafayette, IN, 1995.
CTIC, *Leading and Communicating, a Guide for Watershed Partnerships,* Know Your Watershed Kit Guides, CTIC, West Lafayette, IN, 1995.

REFERENCES

1. Walters, B., Organizing a watershed alliance, *Land and Water*, November/December 2000, pp. 43–44.
2. Conservation Technology Information Center (CTIC), *Putting Together a Watershed Plan, a Guide for Watershed Partnerships*, CTIC, West Lafayette, IN, 1995, p. 16.
3. Holdren, C., Jones, W., and Taggart, J., *Managing Lakes and Reservoirs*, N. Am. Lake Manage. Soc. and Terrene Inst., Madison, WI, 2001.
4. Norbeck, C., Is the watershed approach here to stay? *Natural News*, EPA 908-R-99–006, USEPA, Denver, 1999.
5. Born, S. and Genskow, K.D., Exploring the Watershed Approach: Critical Dimensions of State-Local Partnerships, the Four Corners Watershed Innovators Initiative, final report, River Network, Washington, D.C., 1999.
6. Birchford, S.L. and Smolen, M.D., *A Manager's Guide to NPS Implementation Projects*, NCSU Water Quality Group, Raleigh, NC, 1990.
7. New Jersey Department of Environmental Protection (NJDEP), *Planning for Clean Water, the Municipal Guide*, NJDEP, Trenton, 2000.
8. CTIC, *Building Local Partnerships, a Guide for Watershed Partnerships*, CTIC, West Lafayette, IN, 1995.
9. CTIC, *Managing Conflict, a Guide for Watershed Partnerships*, CTIC, West Lafayette, IN, 1995.
10. ICMA (International City/County Management Association), *Protecting Wetlands, Managing Watersheds, Local Government Case Studies*, ICMA, Washington, D.C., 1999.
11. Urban Harbors Institute, Five perspectives on participation in watershed management planning by local governmental officials, *Coastlines, Information about Estuaries and Near Coastal Waters*, Issue 105, October 2000.
12. Riley, A., *Restoring Streams in Cities, A Guide for Planners, Policymakers, and Citizens*, Island Press, Washington, D.C., 1998.

5 Assessment and Problem Identification Phase

It is in the field where we can find out whether our ideas our applicable, where we find out what the various conditions are that we have to deal with and where we find out what the desired improvements are.

— Dr. H.A. Einstein

5.1 INTRODUCTION

The assessment and problem identification phase helps the partnership and stakeholders find out what is happening. An objective of this phase is to provide stakeholders a grasp of the environmental setting and a qualitative understanding of watershed conditions and problems. The partnership oversees the uncovering of concerns, gathering and analyzing information and data, documenting data and management decisions, and developing an overall goal. It starts with an inventory. Assessment means evaluating available data and information to describe the resource condition and then making a judgment on that condition.

Watershed assessment involves a careful analysis of all water resources in the drainage basin and their stressors to detect what parts of the whole watershed are in trouble and which are contributing to these problems. Watersheds need to be thought of as complex systems with interacting variables. For example, groundwater and surface water are interconnected. In fact, completely separating the two is often difficult because they "feed" each other depending on the hydrologic condition. Impacts on one component of a watershed can have profound effects on other components — the idea that everything is connected to everything else lies at the heart of this watershed planning and management approach. Problem identification entails comparing the assessment judgments to criteria, such as water-quality standards or soil productivity measures such as "T," to detect an impairment or threat to a resource. A complete watershed assessment is needed to start the process to determine the problems and opportunities. For some watersheds, this means an extensive inventory and monitoring effort, whereas in other watersheds, information gaps only need to be filled. Occasionally, assessment of an entire watershed is necessary to determine the appropriateness of various management alternatives for the specific documented conditions. This level of assessment also provides useful information for developing preliminary designs for various management approaches and in developing performance measures with which to evaluate the overall effectiveness of the implementation strategy. The watershed assessment establishes the baseline condition for the land and water resources and provides the basis for future evaluations.

The watershed partnership needs to identify and address concerns about the water and other natural resources, local economy, and social structure within the watershed. Some concerns will be based on perceptions, and others will be based on science. The number of concerns in the watershed is potentially unlimited. Since it is extremely difficult to separate perceptual from scientific concerns, the planning committee must evaluate all concerns. Scoping the overall effort is addressed during this phase with the identification of resource concerns and an evaluation of their environmental impacts. The scoping process means the planning committee (TAC and inventory teams) must base its inventory and evaluation work on the concerns that are most significant to stakeholders. The final watershed management plan explains why certain resource issues were identified as the most significant to stakeholders and why others were not addressed. Local ownership is essential for successful watershed management planning; solving local problems is a key to developing local ownership. Without local ownership, watershed management plans risk never being finished, being shelved and forgotten, or having community members oppose them.

Until the links between the limiting factors and loss of watershed functions can be reasonably demonstrated, it might be impossible to get agreement on actions needed to protect and restore a particular waterbody. This is why gathering, organizing, and publicly reviewing information and data are essential steps in watershed management planning. It is important for the partnership to recognize that there may never be enough data or information about a particular problem or about the interaction of the problems in the watershed to convince *all* the stakeholders that a problem exists. Unfortunately, the perception exists that the more data available to a watershed partnership, the better. Watershed management partnerships cannot continue to assess forever, and each partnership needs to decide on the tradeoff between proceeding with the process and stakeholder comfort level with available information. During the plan development process, the stakeholders and community must prioritize problems and potential solutions, using available information about management alternatives, to form the basis for the watershed management plan. The plan needs to be based on an easy–to-understand water-quality and natural resource assessment, with maps and supporting data.

The partnership must decide how, what, and when information will be available to the public. Information must be available in a timely manner to ensure adequate public involvement. In order to build trust, the partnership must be able to provide interested parties all the information collected in the watershed's assessment, when requested. The planning committee should use information collected in the assessment process to develop a watershed resource library and encourage stakeholders and partners to use it and contribute information to it. A central location for information allows everyone involved or interested in the project access. Partnerships can make the public aware of the availability of information utilizing existing outreach vehicles such as the public water supplies report card on quality, web sites, public libraries, and schools. The partnership's communications plan should outline the specifics of how information and data will be provided to the public.

5.2 ASSESSMENT

Problems and data gaps are identified to set the stage for the development of the watershed management plan during the assessment phase. The outcome of the assessment effort is an evaluation of problem categories, their geographic distribution, and causes. The assessment effort will occur at multiple spatial scales (county and regional issues, watershed and subwatershed issues, site-specific), for multiple sources (land-use pollution, point source, atmospheric deposition) and multiple pollutants (nutrients, sediment). The multiple-scale and source assessments are needed to ensure that the problems addressed through the watershed management process are appropriate for the level of management proposed and that problems are not masked due the scale of the analysis, issue, or potential source of pollutants. Figure 5.1 shows the increase in the level of detail and the scale of analysis. When determining the level of detail for the analysis, consider data availability, time scale, spatial scale, delivery mechanisms, land-use types, management activities, value of resource, and management cost. This level of analysis can depict multiple ecological and environmental impacts such as hydraulic perturbations; stream stability; water

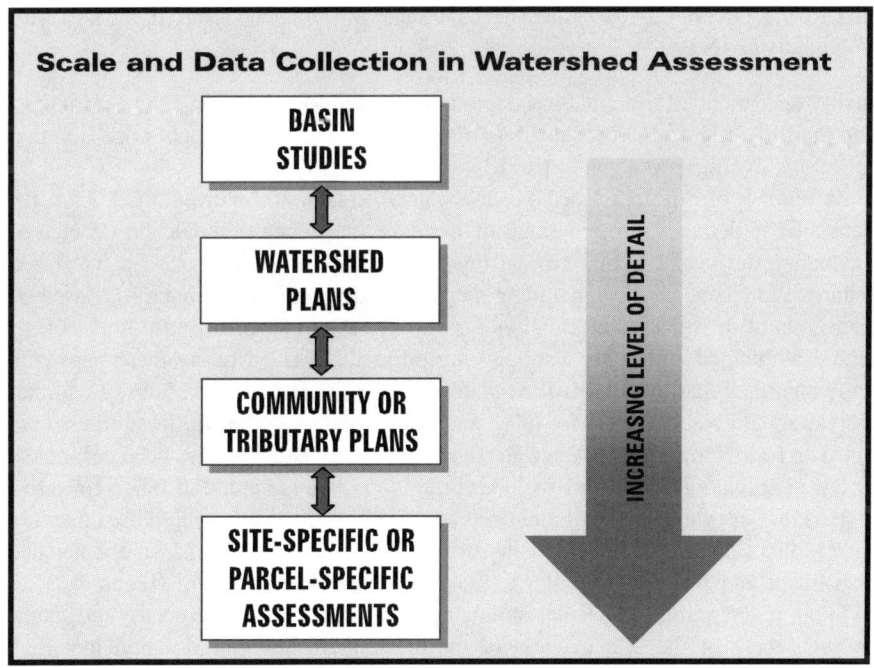

FIGURE 5.1 Scale and data collection in watershed assessment. (Courtesy of CTIC, West Lafayette, IN.)

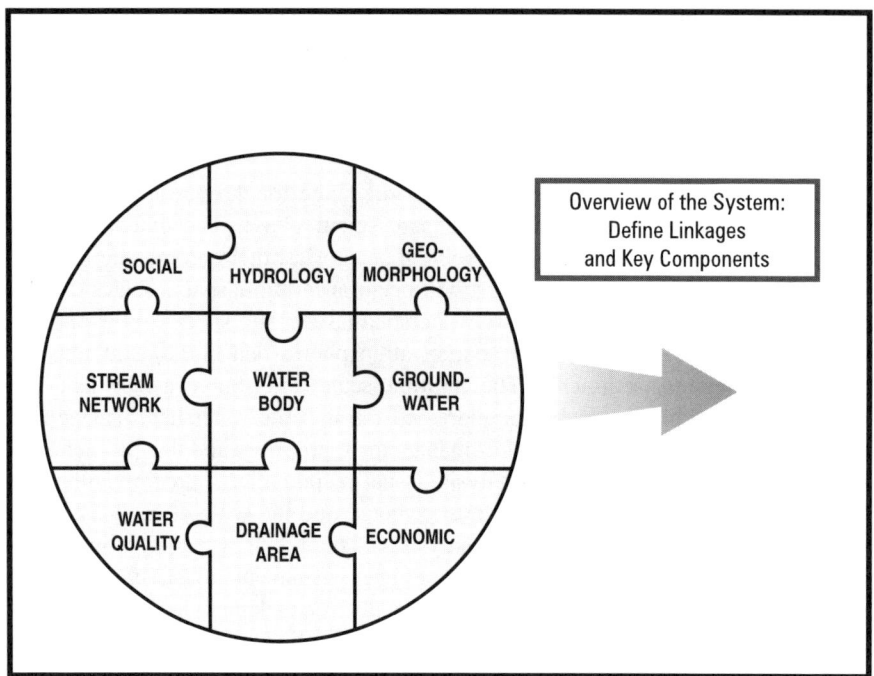

FIGURE 5.2 Overview of the system: define linkages and key components. (Courtesy CTIC, West Lafayette, IN.)

quality and intended uses; habitat degradation; and biological and living resources. The masking effect could result in an emphasis for the management approach that does not solve the problems at hand.

A number of different aspects of the watershed need to be inventoried. After the watershed is defined, the first stage of the assessment phase should be devoted to conducting an inventory and critical review of all existing databases on watershed features and characteristics, including the waterbodies and their tributaries. Based on the results of the review, information gaps on current water-quality use impairments should be defined and either a monitoring effort focused on filling these gaps or a study identified that would provide the necessary information in the future. A similar process should be conducted for other natural resources of concern. Desired uses are based on factors important to the watershed community and what has been designated by law or regulation. They may include current or potential natural resource concerns, such as loss of prime farmland and open space. Although many desired uses may not have a direct impact on water quality, they need to be considered in the watershed assessment and planning in order to establish a baseline for them (Figure 5.2). In addition to determining the water and natural resources' problems, needs, and status, an assessment of the effectiveness of the existing institutional situation including available programs and their implementation status and political support needs to occur.

The purpose of evaluating the existing projects, programs, and ordinances in the watershed is to determine what is already being done and what can be improved upon to restore and protect water quality. Once opportunities are identified and the gaps have been resolved, strategies can be developed to include them in the watershed management plan.

5.2.1 DEFINING THE WATERSHED

The first step is to determine the scope of the watershed management effort and then map it. The boundaries of the area under consideration need to be defined. Most successful partnerships work with a manageable-size watershed, yet it encompasses all different, but integrated, areas. The geographic boundary provides a spatial context for assessment and a sense of place for the partnership. An established watershed boundary streamlines the process of gathering, organizing, and presenting information for decision making. When the boundaries are set, the area within should reflect relevant hydrologic processes. The watershed encompasses not only the water resources, such as a wetland, lake, aquifer, or coastal zone, but also all the land that drains into the resource. Analogously, a subwatershed is the drainage area of a tributary of a water resource. The distinction between a watershed and a subwatershed is one of scale and nomenclature. A drainage area of a tributary is both the watershed of the tributary and a subwatershed of the main waterbody. For assessment and management purposes, designating subwatersheds during this phase is helpful.

A topographic map is the most commonly used tool to define the watershed boundary. While defining a watershed boundary using a topographic map is quite easy, and following the basic principle that water runs downhill, it should be viewed as preliminary until groundwater–surface dynamics have been identified. Occasionally the TAC can run into difficulty combining boundaries of surface and recharge areas. If this occurs, the TAC needs to consider combining surface and groundwater into a single, larger area. The frequent interchange between surface and groundwater ensures that what pollutes one pollutes the other. Delineating a watershed manually is not as simple as it sounds and requires field checking. A procedure for delineating a watershed boundary is outlined in a four-page fact sheet entitled "Delineating Watersheds: A First Step Towards Effective Management"[1] and in the *Texas Water Watch Manual* for conducting a watershed land-use survey.[2] In addition, a number of digital terrain model (DTM) computer programs can be used to establish surface watershed/subwatershed boundaries on topographic data layers.

Where groundwater is a source of drinking water within a watershed, the community must identify the wellhead area and other relevant groundwater recharge areas. Using different methods can significantly affect how a groundwater wellhead/recharge area will appear on the assessment mapping and the watershed management planning process. There are three different methods to delineate wellhead areas:

1. Arbitrary fixed-radius method. Concentric circles are drawn around a well to identify where contamination would most likely reach the well. The method requires little technical expertise and is fast and inexpensive. However, it does not consider physical/geological features or groundwater movement.
2. Calculated fixed-radius method. A more precise delineation method that considers groundwater flow, soil conditions, geology, and the physical

processes of contamination. This method identifies "time of travel zones" to represent the amount of time it would take for a contaminant to reach the well if the water withdrawals remain constant with current levels.
3. Analytical method delineations. This approach uses computer models and a variety of data points to delineate the source of groundwater drinking water. This approach is the most expensive and requires a qualified professional to complete.

A watershed approach to planning requires that boundaries for examining the relationships between the natural environment and human activities be based on the biophysical and not the political boundaries. No formula exists for determining the appropriate geographic scale for any particular watershed effort. The effective size of a watershed is influenced by drainage patterns, stream order, stream permanence, climate, homogeneity of land uses, watershed geology, and geomorphology. Each factor is important because each influences stream characteristics, although no direct relationship may exist. Two examples from Michigan show the wide range in watershed size based on the identification of the geographic scale of the impairments. The Boardman River covers 295 square miles. The planning group identified one pollutant of concern, sediment. The primary pollutant source was stream-bank erosion, and since the predominant land use was forestry, the whole watershed was considered manageable as a project size. Usually it would be considered too large to address as a single project and would be addressed on a tributary-by-tributary basis like Davis Creek. When selecting Davis Creek, the local partnership decided to focus on one small tributary to the Kalamazoo River. The small size of the watershed, approximately 16 square miles, made it manageable to address agricultural pollution in the headwaters and urban issues in the remainder of the watershed.[3] The Illinois EPA[4] recommends watershed planning efforts be initiated on a watershed scale of 50,000 acres (78 mi^2) or less because of the amount of detailed information required in its watershed resource inventory.

Much of the existing information and data about natural resources are organized along political boundaries such as counties, states, and school districts. This information needs to be reorganized along hydrologic boundaries. Generally, three basic maps are needed to provide a foundation to organize watershed information: a county-level highway map, U.S. Department of Interior Geological Survey topographical map, and the USDA soil surveys. The topographic and soil survey maps are used to:

Identify the watershed and subwatershed boundaries using the topographic map.
Identify waterbodies, groundwater recharge areas, and aquifers.

A basic starting point in developing a watershed plan is developing the base map, which builds off the combined characteristics of the highway, topographic, and soil survey maps. At a minimum the base map should include the following information: watershed (subwatershed) boundaries, local political jurisdictions within the watershed, the stream network, existing infrastructure and infrastructure plans, an inventory of the existing land uses in the watershed, mapping and its natural features,

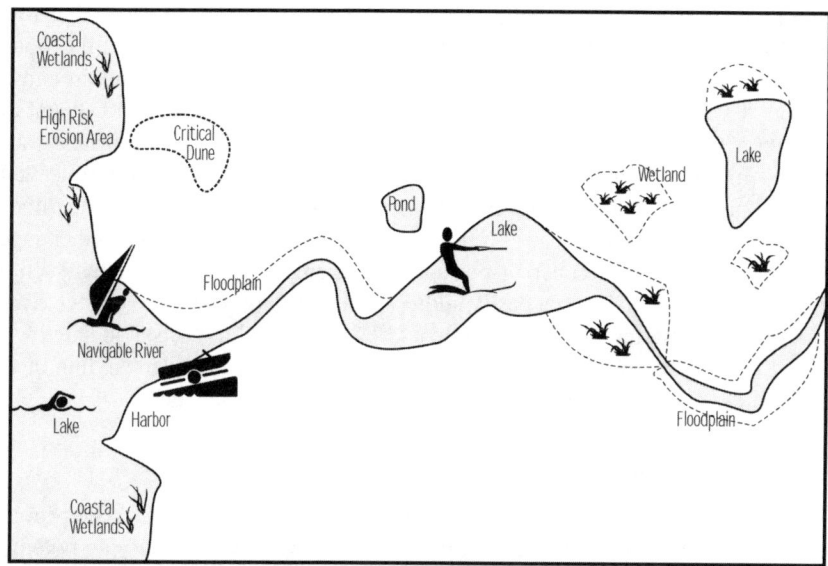

FIGURE 5.3 Highlights the mapping of resource areas in a watershed. (Reprinted with permission from Terrene Institute, Alexandria, VA, as published in *A Watershed Approach to Urban Runoff; Handbook for Decision Makers*.)

and government-protected areas (Figure 5.3). The watershed features map represents the physical features of the land such as mountains, valleys, ridges, lakes, wetlands, groundwater recharge areas, and stream channels. The mapping should include areas that are subject to natural hazards such as flooding, wave action, earthquakes, and landslides. These natural hazard areas are highly correlated with geologically vulnerable areas, such as shallow soils overlaying fractured limestone or bedrock and steep slopes. Generally, the steeper the slope, the greater the erosion hazard is.

In addition to topographic maps, aerial photographs are essential for watershed management projects. Aerial photos provide a feel for how the watershed is connected. Stream-bank erosion rates can be identified from sequential aerial photographs on larger rivers. Lehre[5] and others estimate sediment production rates by using time-trend aerial photography to determine lateral erosion rates and bank heights determined in the field. Historical maps and photographs provide important information on the how the watershed and drainage system appeared previously. This is important when looking at restoration opportunities. Photographs provide a visual image to illustrate problems that need to be resolved or corrected. They generate interest in the watershed, and photos are inexpensive for the public education benefits they provide. Satellite imagery and thermatic (land use and vegetation) mapping are useful resources in determining baseline conditions on a watershed basis.

5.2.2 Project Scope

Once the watershed boundary has been delineated, the watershed needs to be examined using a four-dimensional framework. A lateral assessment of the watershed needs to be made to document the movement of water, energy, and material from the upland areas into the drainage network. A longitudinal assessment of the watershed needs to identify the different stages of the water resource continuum and its condition. A vertical assessment of the watershed needs to be made to document the movement of water, energy, and material into and from groundwater to surface-water resources. In addition to defining the watershed's spatial boundary, the time scale most appropriate for planning and management must be identified. The time dimension is critically important because waterbodies are constantly changing and changes can be detected in any number of periods. For example, legacy problems such as contaminated sediments are not easily addressed and usually require extensive remediation over the long-term as compared to addressing annual impacts associated with changing crop rotations. These scales of assessment require an understanding of the biological, chemical, and physical processes that define a watershed.

5.2.3 Inventory Stage

Inventory of the available data is a key step in the assessment process. The primary uses of the inventory include screening for pollution problems, identifying potential sources of pollution, helping interpret the baseline for possible restoration efforts, establishing a sense of value for the stakeholders, and providing an educational framework for the public and schools. The planning committee needs to create a team to oversee the inventory effort. The inventory team should include anyone who has knowledge of local resources and a desire to help. Team members should be added as the need arises. Local colleges and universities may be able to help with inventories and surveys. Local citizens have a vital role to play as volunteers can carry out many of the inventory responsibilities. Visual inventories can provide a great opportunity to involve the steering committee in the project and familiarize the members with the watershed. In the Iowa RCWP, heavy sediment and a blanket of corn stalks covering a recreational lake surrounded by farmland helped make the problem and its source especially clear.[6] The visual inventory should be a periodic look at the watershed and incorporated into the ongoing monitoring effort.

Four types of data are useful to the watershed planning committee: abiotic, living, social, and physical land features/uses. Abiotic data are water discharge, weather, and temperature. Living data are related to flora and fauna resources. Social data are demographics and value information of watershed residents and other stakeholders. Physical features and land-use characteristics data establish the physical constraints for management. A number of actions are completed during this phase: collection and evaluation of existing data, an overall characterization of the watershed, and identification of critical areas. Lack of data and uncertainty about specific issues are common to watershed planning. A *data gap* occurs when not enough information exists to make an informed decision about a resource's condition or its limiting factors. Filling the data gaps is key to improving decision making, and strategies for obtaining the necessary data should be included in the watershed management plan. In order for a partnership to think productively about the problems of managing its watershed, it needs to distinguish among the concepts of data, information, and knowledge. The partnership needs to view data as raw or unabridged descriptions, observations, and measurements of watershed factors, and information as patterns that the analyst finds or infuses in data. Knowledge is the product of translating data, information, ideas, and experience. A managing principle for a watershed partnership is to maximize the use of and access to the individual partners' knowledge.

The goal for the data management effort is to enhance partnership performance by explicitly designing and implementing tools, processes, systems, structures, and cultures to improve the creation, sharing, and use of all types of water-quality and natural resource data. Several key concerns need to be addressed when examining data and information:

- Appropriateness of data and information. Is the collected data and information what was needed?
- Frequency of collection. Does the frequency of collection match the analysis requirements?
- Data and information reliability. Do the data and information meet QA/QC requirements?
- Cost of obtaining the data and information. Is it worth it?

What facts do we need to know about the watershed if we are going to achieve our view of the watershed's desired state? What data is needed to identify natural resource, social, and economic trends? How much data do we need in order to feel comfortable in accepting the condition of the watershed? What environmental indicators can be used?

The planning committee needs to conduct an inventory of available information on point and nonpoint sources using information available from federal, state, and local agencies and other organizations' databases. Information about past, present, and projected economic and social conditions in the watershed is also needed to make a comprehensive evaluation of the watershed. Inventory institutional and management features of the watershed; this will provide a social and political basis on which to build programs and strategies. A few examples are waste disposal, local health department programs, stormwater policies and programs, zoning plans, vegetation management programs, shorelines protection and management programs, as well as the development/redevelopment policies and programs. The CWP[7] listed 34 different data layers and their importance for management approaches. The TAC must decide what data and information are needed based on the mission statement and its watershed knowledge. Data of both the historical and existing conditions related to watershed structure and functions as well as social, cultural, and economic values are important. Methods used for understanding the socioeconomics of watershed residents will depend on the watershed communities and on the resources available to collect, analyze, and interpret the information. The two main categories of information about communities and their residents are primary data and secondary data. Primary data include first-hand interviews, review of newspapers, focus groups, and citizen surveys. Secondary data include population, housing, economic, and agricultural census data that various agencies collect at the national, state, and local levels. The easiest and most effective way to learn about communities is to check the census (secondary data source) and talk with people who live in the watershed (primary data source).

Identifying temporal/seasonal issues affecting things such as discharge rates, receiving water flows, and designated or existing use impacts is necessary in order to define critical periods for management purposes. Identify and document all ongoing watershed efforts and volunteer monitoring programs in the watershed. Watershed management plans cover all activities within the watershed that have an impact on its natural resources. Information related to ground- and surface waters, cultural resources, threatened and endangered species, and laws and local regulations is usually included in a watershed inventory. Each partnership needs to collect resource information specific to its situation, identified concerns, and desired future condition.

Watershed management plans typically include the following information: hydrology, rainfall characteristics, topography, soil types, water and natural resources, land ownership, environmental monitoring and assessment stations, land use, and point sources of pollution (landfills, POTWs, underground tanks, etc.). The plans usually reference county soil surveys, storm sewer maps, aquifer maps, Section 305(b) and water-quality assessment reports, diagnostic and feasibility studies, community water supplies, fish populations, agricultural statistics yearbook, attitude trend surveys, county comprehensive plans, census data, and chambers of commerce and economic growth studies. Newer plans include imperviousness area estimates. This is an appropriate indicator with which to measure the probable impacts of land development on aquatic systems and to project future impacts.

Watershed inventories should focus on the best information available at the time; as new data become available, they should be added to the inventory and the watershed plan should be adjusted accordingly. Rarely is a resource inventory an exhaustive study of the system. However, if the inventory can clarify resource concerns, help local people understand cause-and-effect relationships, and set a course to reach a desired future condition, it has achieved its goal. The "Inventory and Evaluation Procedure for Land Treatment"[8] was developed and utilized to alleviate concerns associated with agricultural lands. This watershed screening-level procedure identifies problem agricultural land and the assistance needed at three possible levels of intensity: minimal, moderate, and major. This level of information allows the watershed planning committee to work with stakeholders to develop management approaches to implement solutions.

The information collected by the inventory team should be presented to the watershed partnership via meetings and reports. The team should be completely nonjudgmental in its presentation and use maps, tables, charts, and visual tools to make the information easier to digest. The team should not try to draw conclusions from information, as the inventory team does not make decisions for the watershed partnership. The team should prepare a written report detailing what they did and what they discovered; this report will become a part of the watershed plan.

The next phase of the initial inventory is to identify the known water-quality and quantity concerns. A central objective of watershed management is to maintain the integrity of the major hydrological pathways that people can affect. A hydrological analysis can range from a simple review of high-flow data, to a detailed survey

Assessment and Problem Identification Phase 71

of flow fluctuations and base stormwater controls. The type of hydrological analysis needed will be defined in the context of the issues and concerns investigated. Therefore, the assessment report needs to focus on how people have changed the landscape in a way that introduces materials and contaminants into the hydrological pathways — and thus affect the amount, timing, and quality of water. Before water quality can be improved, the causes of current problems and issues must be known. Collection of long-term trend data on factors such as stream flow volumes, water quality, fish and wildlife habitat characterization, and population information is key to determining the conditions and trends in a watershed. Unfortunately, most watersheds lack basic data or baseline data on the conditions of the water resources within them, and ongoing trend data are becoming harder to find as a result of declining funding in the area of monitoring.

From a planning perspective, there is always a discrepancy between database size, data needs, and data use. The size of the database has nothing to do with the utility of the observations and measurements for their use in an assessment. It has been estimated that over 50% of the collected information never gets used. Partnerships need to adopt an outcome approach to their data management efforts, such as data need to have an actual, verifiable purpose before collecting; data must be able to connect the management efforts with environmental results; and data must provide the basis to help partnerships improve their performance. For watershed analysis, collected data need to have three competencies: understandable, helpful, and useful. These three competencies are mandatory for watershed management efforts to be successful.

Budget and technical limitations make it impractical to collect all available information and fill all data gaps. Therefore, prioritizing the data needs is an important task for the TAC, planning, and steering committees. At a minimum, the data necessary to explain the mechanisms and processes that affect water-quality conditions need to be collected.

5.3 ANALYSIS

The condition of a waterbody at any one time is the result of a complex interaction of many different physical, chemical, and biological factors. The greatest influence on these variables is human impact on the watershed. Rainfall cycles, watershed characteristics, and waterbody characteristics all contribute to the existing condition. The physical and chemical factors in turn support a community of biological organisms unique to the watershed. Understanding the fundamental ecosystem processes is critical to managing a watershed effectively. One of the most popular techniques is to use a mass balance approach. An example of the value of utilizing such an approach is best illustrated by the following example from the White Clay Lake Project. Originally, White Clay Lake, located in Wisconsin, was assumed to be a surface-water-fed lake. The water budget identified that surface runoff contributed only 35% of the water volume, groundwater was 40%, and atmospheric contribution was estimated to be 25%. This implies that watershed management might not have significant influence on in-lake water quality due to the small percentage of the total

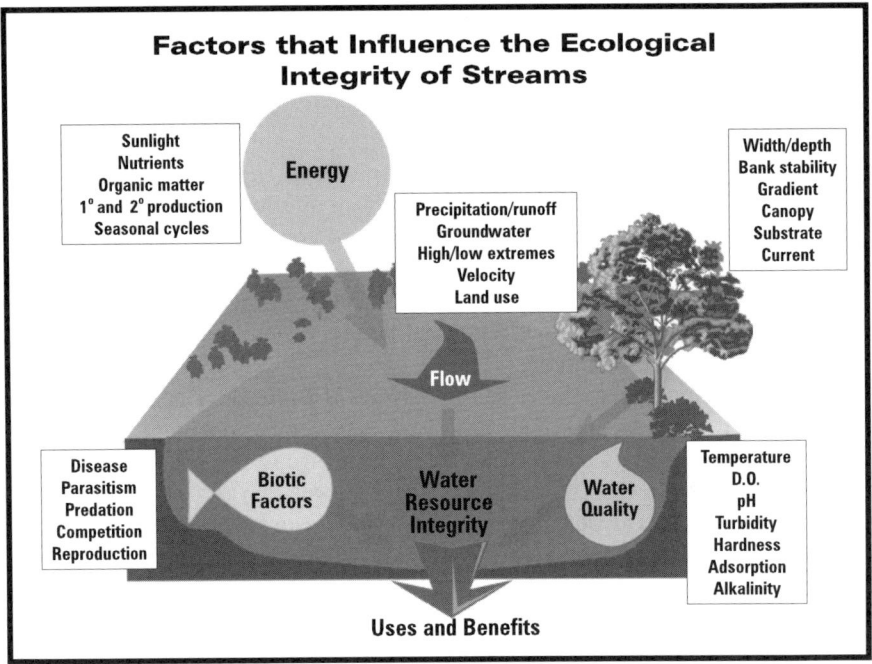

FIGURE 5.4 Factors that influence the ecological integrity of streams. (Courtesy of CTIC, West Lafayette, IN.)

volume of water being surface runoff. While the water budget is important, the pollutant budget had to be developed to get a clearer understanding. For White Clay Lake the phosphorus budget indicated that over 50% of the total load was associated with surface runoff, so the possibility of watershed management's reducing phosphorus loading to the lake and thus improving water quality existed.[9]

Evaluation of resource quality and condition requires an assessment of waterbody quality and condition and is probably best conducted using a methodology that evaluates the condition of the biological community. The functioning of many waterbody uses is directly related to the biological integrity, since the biota reflect the overall health of the system (Figure 5.4). Therefore, an assessment of the condition of a waterbody can then be based on an evaluation of the relative "biotic impoverishment." Assessment of resource conditions must be independent of resource value. Resource value should be broken into various components. Although potentially contentious, such value analysis would enhance the objectivity of the assessment process and clarify the thinking of the assessors. For example, take the watershed's fishery resource. The value of certain species might basis the analysis of the overall condition of the fishery resource and promote an ecologically deficient system. The purpose of assessing current conditions or ecological integrity is to evaluate the overall health of the aquatic resource, to determine if a designated use is being attained, and to evaluate the ecological potential of the water resource. A number of rapid bioassessment techniques exist, such as the USEPA's Rapid Bio-

Assessment and Problem Identification Phase

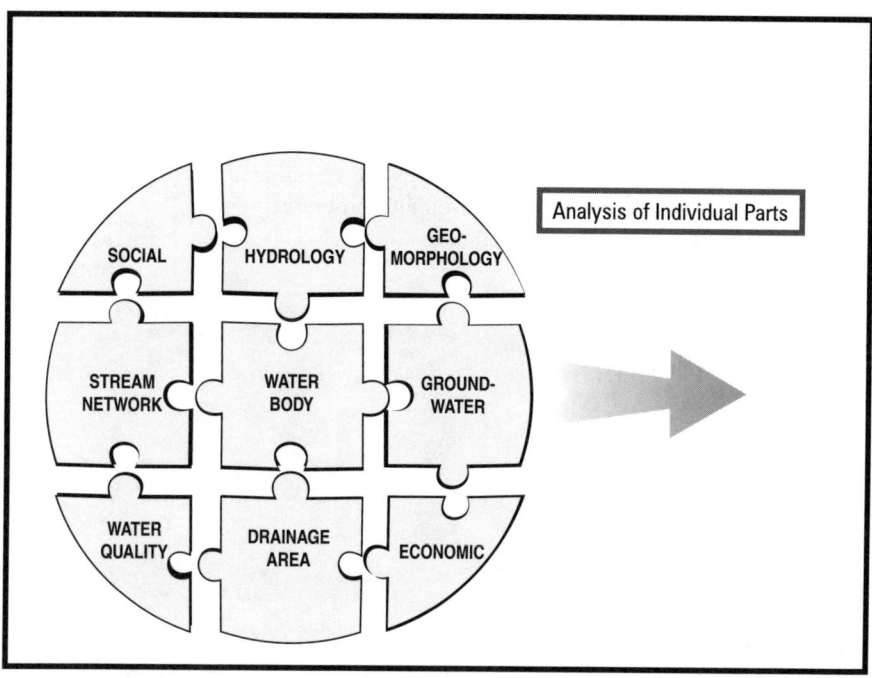

FIGURE 5.5 Analysis of individual parts. (Courtesy of CTIC, West Lafayette, IN.)

assessment for Use in Streams and Rivers[10] for collecting and integrating habitat, water-quality, and biosurvey data to evaluate current conditions. Habitat is an important determinant of ecological potential and provides the basis for further ecological investigations.

The watershed assessment can be divided into four components: watershed health, water quality, water quantity, and watershed biology. All these components are interrelated and need to be examined in that context (Figure 5.5). The analysis of the individual components and their interrelationship will decrease the basis and allow for the development of management strategies to address the issues and problems identified by the component. Watershed health is the most complex and wide-ranging component. FISRWG[11] identifies four components (hydrologic and geomorphic processes, chemical and biological community characteristics), which must be integrated to understand stream health, that can be used to establish watershed health. The planning committee and the chairperson of the TAC are responsible for ensuring the management approach for each component is coordinated and consistent.

Once a comprehensive set of data and information from past studies as well as from any current monitoring has been collected and a report prepared on this review, a stakeholder-developed consensus should be formulated on what real water-quality use impairments or threats exist in various parts of the watershed. When these problems have been identified, and if the cause of the problem has not been determined, site-specific studies should be undertaken or identified to determine the cause or specific pollutants responsible for the use impairments or threats. In stakeholder-based watershed management, a designated beneficial use impairment of a waterbody must be perceivable by the public and not based simply on exceeding water-quality standards or objectives. Older water-quality standards may not adequately protect aquatic life or may no longer be reasonably achievable due to human-induced irretrievable impacts. The primary criterion for water quality is whether the waterbody meets designated uses. Designated uses are recognized uses of water established by state, tribal, and federal water-quality programs.

Designated uses for surface waters are:

1. Agriculture
2. Industrial water supply
3. Public water supply
4. Navigation
5. Warm-water fishery
6. Cold water
7. Partial body contact
8. Total body-contact recreation

5.3.1 THREE-TIER ANALYSIS

The three-tier analysis is one method that can be used to help in problem identification (Figure 5.6). This approach will help partnerships avoid the common problem of single issues' driving the planning process. Watershed planning entails reversing the paradigm and offers the opportunity to use a holistic view of the system and a chance to identify all the problems.

- The first tier identifies all the natural resource components, and where they are linked to the watershed problem, perhaps beyond the watershed.
- The second tier identifies the stressors acting to degrade or impair the identified components in tier one. For water, quality limited water resources require identification of the designated use impairment and the pollutant/stressor. Thus if recreational use (swimming) of a lake is impaired, turbidity might be listed as a stressor.
- The third tier is where we look at the sources that contribute to the stressors. For the stressor turbidity in the lake cited in tier two, soil erosion and algae might be listed as the sources.

The three-tier analysis of the problems in the watershed area will identify all potential aspects of the problems and provide a starting point to prioritize as well as indicate what additional information and data are needed. An important characteristic of the recommended model is that with inventories and problem identification completed, credibility is established by simultaneously correcting identified problems. New and more accurate information can be incorporated at each step, depending on qualitative and quantitative information and flexibility in analysis. The three-tier analysis allows the partnership to answer the following questions:

Assessment and Problem Identification Phase

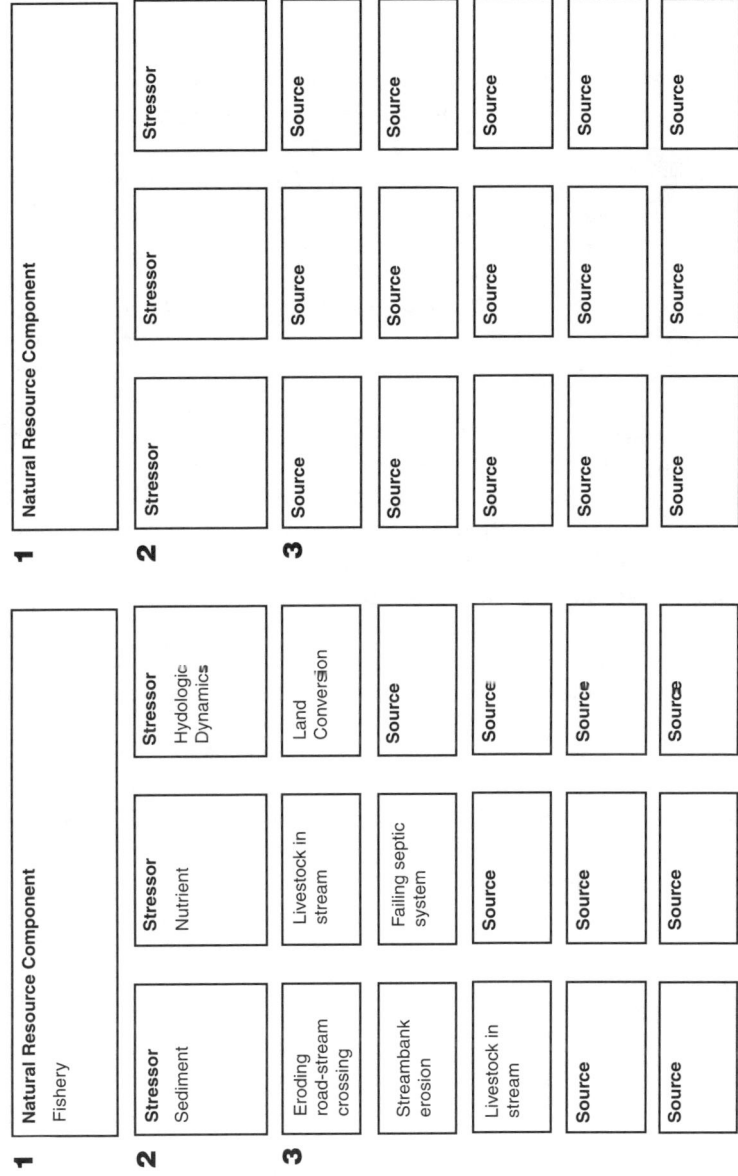

FIGURE 5.6 Three-tier analysis process.

- How do these stress/disturbances relate to the watershed processes, structure, and function?
- Are the disturbances physical, chemical, or biological?
- What is their origin?
- How will the size and characteristics change given current and future conditions?

Stressor identification approach leads professionals through a rigorous process to identify stressors that cause biological impairment in aquatic ecosystems and assemble cogent scientific evidence supporting conclusions about potential causes. Accurate identification of stressors causing biological impairment will help develop management actions to protect or restore watershed. The use of stressor identification is prompted by a change in biological assessment data for a given watershed indicating that a biological impairment has occurred. The general process entails critically reviewing information, forming possible stressor scenarios that might explain the impairment, analyzing those scenarios, and producing conclusions about which stressors are causing impairment. The level of detail required varies by severity and type of biological impairment. The process is iterative, usually beginning with retrospective analysis of available data. The accuracy of the identification depends on the quality of data and other information used in the process. In some cases, additional data collection may be necessary to identify the stressors accurately; for example, sediment versus turbidity, nitrogen versus phosphorus, and phosphorus versus sediment. The conclusions are then translated into management actions through the watershed plan development process. The effectiveness of those management actions can be monitored against changes in the resource condition or a reduction for pollutants targeted for reduction.

When properly applied, environmental indicators provide valuable insight to watershed problems and can be used to support the management and decision-making process effectively. However, indicators are often based on limited data; thus, prudent choices must be made in their selection, application, and interpretation. The partnership must ensure the indicators are relevant to the watershed management plan and target appropriate activities. For example, for issues related to the adverse impacts of soil erosion problems, the on-site indicator should be soil erosion rates and the waterbody indicator should be suspended sediment concentrations (SSC) rather than total suspended solids (TSS) or turbidity. SSC is more reflective of in-stream relationship with soil loss, but TSS is comprised of sediment as well as organic particles. Turbidity has similar problems.

Watersheds are highly interactive systems; understanding how water flows into and through the stream network is critical to watershed management. It is impossible to alter one factor or characteristic without affecting another part of the system. Once the water-quality conditions have been identified and the causes of these conditions described, the key remaining question is whether the causative factors are a function of and responsive to management. The ability to identify influences is important for identifying alternatives during the planning phase. The assessment phase includes a description of past management influences to prevent the repetition of previous mistakes and should facilitate prediction of future watershed response

Assessment and Problem Identification Phase

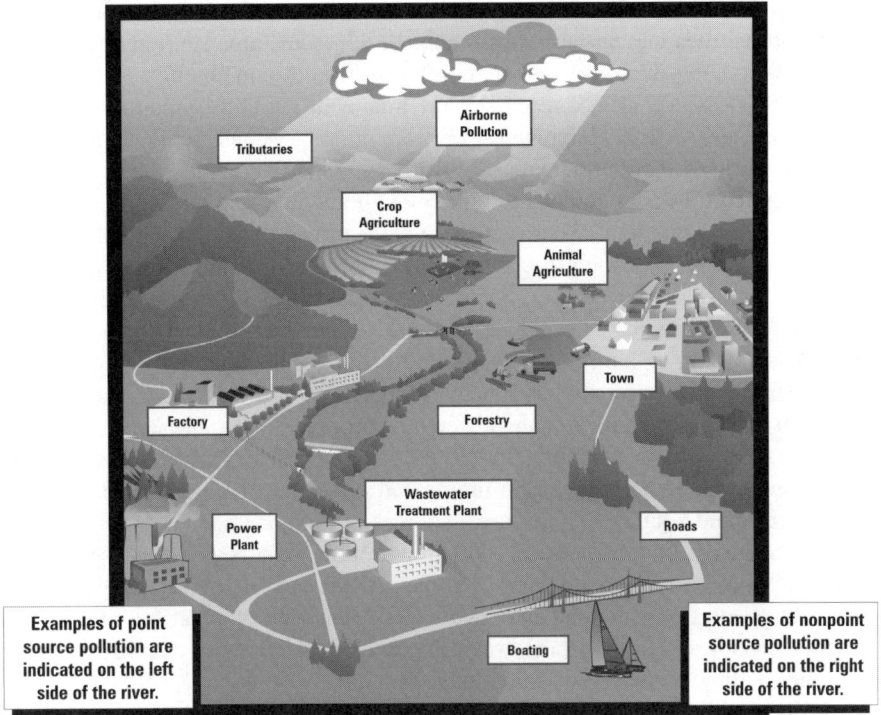

FIGURE 5.7 Various pollutant sources in a watershed. (Courtesy of CTIC, West Lafayette, IN.)

for evaluating alternatives. Recognition of management influences also is important for predicting the effectiveness and feasibility of specific management approaches.

5.4 POLLUTANT SOURCE ASSESSMENT

Water pollution caused by human activities is usually from either point sources or nonpoint sources (Figure 5.7). These terms indicate how pollutants are generated and released to surface and groundwater. Other sources that are not classified under point or nonpoint sources include underground petroleum storage systems and many large and small businesses like dry cleaners, restaurants, and automotive repair shops.[12] In the U.S., approximately 2000 sewage treatment plants and industrial facilities discharge effluent, treated to various levels, directly into the nation's waters. Municipal sewage in the U.S. is treated to meet secondary treatment standards or more stringent water-quality-based limits prior to disposal. Industrial and municipal discharges are regulated through permits under the CWA's National Pollutant Discharge Elimination System (NPDES). Permits establish pollution limits and specify monitoring and reporting requirements. More than 40% of the water used in the U.S. is for industrial purposes. Typically about 20% of the industrial use is in the finished product; the remainder is treated and discharged to receiving waters. Municipal discharges come from publicly owned treatment works that discharge into surface waters. About 2.3 trillion gallons of effluent are discharged from sewage treatment facilities into surface waters annually. In some areas during heavy rains, storm sewers and sanitary sewers are combined, resulting in combined sewer overflows (CSOs), which do not reach sewage treatment facilities, going directly into receiving waters instead. Combined sewers are no longer constructed but are still in operation in many older urban areas. In April 1994 the EPA issued the CSO policy. The policy requires communities to implement immediate and long-term actions to address overflow problems. These actions include proper operation and maintenance of sewer systems and capital projects to separate or treat CSOs, as well as public notice in the event of overflows. Separate sanitary sewer systems also experience overflows called a sanitary sewer overflow (SSO) under wet-weather conditions. The USEPA is developing a proposed SSO rule to address this issue.

Over 60% of the remaining sources of pollution are from nonpoint sources. Nonpoint sources come from many different sources and enter surface and groundwater in several ways. Nonpoint source pollutants can enter waterbodies through direct runoff, runoff through storm drains and sewers, wet- or dry-air deposition, and underground aquifers. The primary sources of nonpoint source water pollution in the U.S. are agricultural activities and urban runoff. Pollutants include sediments, nutrients, pesticides, and pathogens. Sediment is the largest source. Assessment of sediment needs to be divided into two main categories of processes: those that generate erosion such as rill, sheet, wash, dry ravel, freeze-thaw; and those associated with the routing of eroded soil to drainage network.

In source analysis for a watershed management plan, the relative contributions of different sources are assessed. An estimate of pollutant loads from both point and nonpoint sources is essential to this analysis as the ability to determine if pollutant loading exceeds water-quality standards. In order to clarify source contributions of nitrogen to the Indian River ecosystem, a loading model was used to assess current nitrogen loading to the lagoon and predict conditions that may result from new development and eventual buildout of the watershed. The model took into account a variety sources, including sewage, fertilizer, road runoff, and precipitation. The results indicate that on-site sewage disposal systems (OSDS) contributed 12% of the nitrogen load to the lagoon, while agricultural lands contributed 31%.[13] The OSDS contribution was roughly equivalent to cattle (13%). This allows planners to identify or focus on the sources with the greatest management potential. The loading model provided a mechanism for calculating the total amounts of nitrogen contributed to a system and predicting the concentration of nitrogen in groundwater based on hydrological parameters. A follow-up study to clarify OSDS contributions to the Lagoon was completed in 2000. The study confirmed the earlier estimates.

Through the planning process, the planning committee will develop the estimated load reductions needed to meet water-quality standards as minimum planning targets against which to evaluate different management approaches. This concept is consistent with establishing a total maximum daily load (TMDL) estimate for a waterbody. The load allocation for nonpoint sources and the wasteload allocation for point sources are determined from an analysis that links the achievement of a water-quality standard to various management alternatives that can be applied to the identified sources.

The types and concentrations of compounds found in certain bodies of water are closely linked to land use in surrounding areas (drainage area). In fact, water in urban settings has a characteristic chemical makeup or signature that is different from runoff found in agricultural settings. The chemical makeup correlates with the chemicals used in the setting or drainage area. For example; total nitrogen concentrations for urban runoff range from 3 to 10 mg/L while agricultural runoff ranges from 0.77 to 5.04 mg/L and total phosphorous concentrations of 0.2 to 1.7 mg/L and 0.085 to 0.1 mg/L, respectively.[11] The long- and short-term environmental impacts associated with urban runoff influence water quality, flow regime, habitat, and energy sources. Water quality is impaired by land uses that contribute pollutants to groundwater and runoff. Flow regimes are affected by increased velocity of runoff, increased volume of runoff, and reduced infiltration into groundwater. Habitats are impaired or destroyed by changes in morphology, condition of banks and upland areas, presence or absence of vegetation and debris, and quality of substrates. Energy sources are reduced when light energy is not being effectively transmitted or when chemical energy in the form of organic materials and nutrients is not present in sufficient quantities to sustain the food web.

An objective of the assessment activities is to identify types, magnitude, and locations of potential sources of pollutant/stress. Many techniques and approaches are available to locate, identify, and map sources. The most common ones focus on modeling nonpoint source land-use contributions, compare loadings for various sources, and examine variability in loadings/concentrations. For point source, locations and magnitudes should be available from the permitting agency. Models or monitoring information is used to identify nonpoint source problems. A number of techniques such as the simple method or USLE are used to calculate loading estimates by source category and timeframe. The estimates are mapped and used in the ranking of pollutant/stressor sources. This type of presentation is important for public education purposes.

Traditionally, an underestimated source of water-quality-related problems is stream-bank instability. For example, in Rock Creeks, Idaho's RCWP project, streambed quality and trout reproductive capacity were reduced by siltation, and transparency was reduced by high suspended sediment concentrations. Initially, at the basin scale, sediment was identified as the pollutant of concern and agricultural activities were identified as the primary cause of impairments. Preliminary analysis showed stream-bank erosion was also a major contributor of sediment loads. The initial project estimates had considered sediment contributions from the two major sources, stream-bank erosion and irrigation return flow, similar in magnitude. However, the 1987–1990 monitoring results indicated that stream-bank erosion contributed two to five times the sediment added from irrigated cropland in the subwatersheds during the May through August irrigation season.[7] The influx of sediment from stream-bank erosion made it difficult to document the effectiveness of the new irrigated cropland management efforts. This new information was utilized to explain the lack of water-quality improvement; the problem of stream-bank erosion continued to mask in-stream benefits from the land treatment. Streams are the lowest point in the watershed of a system and tend to magnify the effects of land-use modifications and practices. Numerous studies have demonstrated that stream-bank erosion contributes a large portion of the annual sediment yield. Over 90% of the total suspended sediment load for West Fork of the Madison River, Montana, was from the stream bank[14] and 80% for Court Creek, Illinois.[15] Stream-bank erosion is a natural process that has two main components: stream-bank characteristics and hydraulic/gravitational forces. Major land-use activities can affect both of these components, leading to accelerated bank erosion. Since bank erosion is often a symptom of a larger, more complex problem, the long-term solutions often involve much more than stream-bank stabilization.

Ideally, a technical team that represents the following disciplines should survey channel segments: engineering, aquatic biology, geomorphology, and soil science. A stream-reach reconnaissance should consist of a combination of physical and ecological reconnaissance techniques that are cost-effective and be able to facilitate comparison among reaches. The combination of techniques should provide sufficient information to facilitate partnership decisions and the options for preservation, control, and mitigation of impairment. Each channel segment should have the following evaluated: general channel stability, locations of ongoing and anticipated bank loss, specific modes of bank failure, bed and bank sediment composition, channel and stream-bank morphology, and riparian vegetation condition. Natural and anthropogenic features that impact channel stability and ecological character are to be located and described. It is particularly important to view time as well as space in the watershed's riparian corridor. Channel evolution can be characterized by six steps, including (1) natural sinuous, (2) channelized, (3) degradation, (4) degradation and widening, (5) aggradation and widening, and (6) quasi-equilibrium. Riley[16] emphasizes that urban streams typically undergo at least two main cycles of adjustment due to urbanization. While determining which cycle the stream is undergoing is useful, it is more important to determine its location in the channel evolution process. Rosgen[17] identifies nine channel evolution sequences that indicate larger possible morphological shifts and their tendency toward a stable endpoint. The assessment process must identify the stage of the channel in the evolution sequence to develop appropriate management strategies.

5.4.1 URBAN

Many Americans live in areas where water quality is already degraded by urban runoff pollution. The problem spreads as urban and suburban areas expand, impairing waters, destroying habitat, and threatening public health locally and downstream. Streams and rivers are often highly visible examples of a degraded environment. The two leading causes of urban runoff pollution are (1) increased impervious cover resulting in greater runoff and volume and velocity, and (2) increased deposition of pollutants.

However, the lack of a common theme makes it difficult to estimate the cumulative impact of existing urban areas and developments within a watershed. Schueler,[18] among others, has documented that impervious cover is likely the best indicator of development, and in some cases watershed imperviousness should be used to classify the quality of streams. In 1994 Schueler proposed that impervious cover can be a unifying theme to guide efforts of the many participants in urban watershed management. For the purposes of watershed management and study, imperviousness includes the sum of the transport and living habitat systems. The transport system consists of roads, driveways, sidewalks, and parking lots. Living habitat, also known as rooftops, consists of the rooftops associated with home, work, play, and shopping. Imperviousness' impact on the environment been documented. The Center for Watershed Protection did an excellent job in summarizing the environmental impacts of imperviousness.[18] The following highlights some of the major impacts:

1. Runoff. As imperviousness increases, runoff volume increases. Between storms, stream flow is reduced. Small streams that once flowed year-round may become intermittent. This is because stream flow is maintained via infiltration of rainfall or snowmelt. While the lack of infiltration results in higher runoff rates, it is also translated into lower dry-weather stream flows due to lack of groundwater recharge.
2. Shape of streams. Increased imperviousness has been linked to widening of stream channels and down cutting of the stream channel. This change in stream shape usually results in loss of pool and riffle sequences and stream canopy.
3. Water quality. Pollutant loads are a direct function of impervious area and increase with imperviousness.
4. Stream warming. The combination of impervious cover that both absorbs and reflects heat, increasing ground and local air temperatures, and the lack of vegetation to offset the effects of solar radiation results in higher headwater stream temperature.
5. Stream biodiversity. Biodiversity decreases when watershed impervious area exceeds 15%.
6. Other urban water resources. Shellfish beds and wetlands show adverse impacts at imperviousness greater than 10%.

Where waterbodies have already been impacted by watershed urbanization, localities can work to ensure the water-quality standards contain goals for maintenance and restoration. The thresholds provide a reasonable foundation for classifying the potential stream quality in a watershed based on the ultimate amount of impervious cover. The two main causes of stormwater pollution are increased impervious cover (greater runoff and volume and speed) and increased deposition of pollutants. One such scheme is outlined below.[19] It divides urban streams into three management categories based on the general relationships between impervious cover and stream quality:

1. Stressed streams (1 to 10% impervious cover)
2. Impacted streams (11 to 25% impervious cover)
3. Degraded streams (26 to 100% impervious cover)

Depending on the resources, scale, and intended use of impervious information, different estimation techniques may be appropriate. The Center for Watershed Protection[20] identifies five basic questions that need to be answered to guide technique selection.

1. What information is currently available? An obvious way to save and time and effort is to focus on resources that are easily available.
2. How much effort/resources can be used? Some accuracy can be sacrificed to meet the constraints of a limited budge and time line. An accurate measure of impervious is useless if there is no funding left in the budget after the measurement is done.
3. What level of accuracy is needed? Ideally, absolute accurate measurements of impervious area would be determined for every watershed. The level of accuracy needs to be balanced against the other tasks for completing the watershed management plan.
4. How important is future forecasting? For planning purposes, forecasting the future impervious cover is just as important as estimating current levels of imperviousness. The two techniques commonly used to forecast future impervious cover are population forecasting and zoning; neither technique is completely accurate. See the Buttermilk Bay example in the plan development chapter.
5. What are the effects of innovative development? Through this technique, credits are assigned to specific land-use decisions based on the nature of the proposed practices.

BOX 5.1

Impervious cover estimates can be useful in projecting the impact of various buildout alternatives. In order for a watershed project to use impervious cover as an indicator, two conditions have to be met: the need for an accurate measurement of current impervious area, and a method to calculate future impervious area associated with development and redevelopment in the watershed. This section provides guidance on selecting the best method, among four, to meet watershed-specific requirements.[21]

Direct measurement: In this approach aerial photography is analyzed to find impervious surfaces. While directly measuring the coverage of impervious area is the most accurate technique for estimating a baseline condition, it is also the most expensive and time-consuming. The date and availability of the aerial photography are two important considerations. In buildout analysis, the baseline is utilized as the point of departure to estimate future impervious coverage and impacts.

Land use: In this approach land-use categories are utilized to estimate imperviousness based on an average or selected impervious rate for each category. The land-use areas are multiplied by the impervious value for each land-use category found within the watershed. This approach is quicker and cheaper and is usually far less accurate. This approach can be valuable for estimating baseline conditions if the data relating land use to impervious area are accurate. In terms of buildout analysis, this approach can be useful since most zoning programs are related to land use. This is the only approach that can be used to accurately estimate the impacts of innovative site design techniques on future imperviousness within a land-use category.

Continued.

Assessment and Problem Identification Phase

Road density: This approach utilizes the road density as an indirect measure of impervious area. While the method is easy to use and requires little data, it is not very accurate. The lack of data on the relationship between road density and total impervious limits the utility of this approach. For example, impervious area associated with medium–density, single-family homes can range from 25 to nearly 60%, depending on the layout of the streets and parking.[18] Additionally, the city of Olympia[22] found that transport-related imperviousness ranged from 63 to 70% of the total imperviousness in 11 residential, multifamily, and commercial areas studied. While it can have some value for use as a preliminary analysis for estimating baseline conditions, it has little value for predicting future impervious area.

Population: This approach correlates population with impervious area. This approach has limited applicability in developed portions of urbanized areas. While population growth is viewed as a more relevant estimate of future growth than zoning projections, it is quite difficult to use to estimate future or current impacts except on a large scale. While the approach is utilized in a number of places, it is not a recommended technique for developing baseline conditions to use in watershed planning. However, in low- to medium-density residential land uses, using the amount of impervious area per person, estimates of future conditions can be made. In high to very high density, additional population results in little or no increase in impervious area since the area is typically considered built out already. Population growth can reflect a rapid decentralization of both people and urban development. A 1998 study documented that a population equivalent to the City of Racine left Milwaukee County between 1970 and 1980. During that same time, the suburban counties grew rapidly, particularly Washington County (67.7% increase) and Waukesha (42.1% increase).[23] Overall, southeastern Wisconsin region added only 10% more people since 1963. The major impacts are related to geographic redistribution. There is twice the number of households, twice as many personal vehicles, and total driving has tripled. In suburban areas, due to the automobile the transport component is greater than the living habitat component. Since 1980 there has been a sharp increase in per-capita vehicle ownership, trips taken, and miles driven, resulting in a dramatic increase in the relative size in the transport component in comparison to the living habitat component. Transport-related impervious cover has a greater hydrological impact on the environment than living habitat imperviousness. Transport-related impervious areas are usually directly connected to storm drain systems, where residential habitat is usually hydrologically disconnected to the storm drain system by lawns.

5.4.2 Agriculture

Water is essential for agricultural production. Water usage and its drainage from agricultural land constitute the leading cause of the water-quality degradation in the U.S.[24] As noted in the watershed processes chapter, the environmental impacts of agricultural activities are (1) associated with erosion and soil degradation, (2) nutrient enrichment of surface water and nitrate contamination of groundwater, and (3) biodegradable organic matter.

The principal agricultural activities are crop (irrigated and nonirrigated) and animal-related production. The assessment process must inventory cropland and determine crop rotations and existing management practices. Animal facilities (confinement and open feedlots) inventory needs to determine species, size (animal units), location, and manure management practices. For both cropland and animal-related production, the management must be inventoried to determine if a detailed assessment is needed to determine management and treatment of the amount and condition of pasture, hayland, orchards, and woodlands.

When analyzing cropland soil loss, planners need to refer to "T." Soil loss equal to "T" will sustain a soil's productivity level but may still be eroding at a level detrimental to water quality. "T" is a value the agricultural community understands and feels comfortable with. Using "T" allows for determining if soil productivity needs to be a natural resource issue for the partnership. To address water quality, the total amount of soil loss needs to be taken into consideration regardless of "T."

5.5 CRITICAL AREAS

All pollutant sources, point and nonpoint sources, must be considered when developing the watershed management plan. The concept behind identifying a critical area is to narrow the geographic scope of the watershed effort and focus management on the part of the watershed contributing pollutants. Focusing on the critical area helps prioritize the concerns and subsequent actions within the watershed. This section focuses on the identification of critical areas in the watershed. There are two types of critical areas: those that impact water quality and those that need management to ensure further degradation does not occur. Delineation of critical areas is important because (1) some lands will not contribute appreciable amounts of pollutants to the waters of concern, and (2) implementation of management practices in all areas of the watershed is cost-prohibitive. Critical areas are used to make the most effective use of limited resources in improving and protecting water quality. Targeting critical areas enables faster measurable progress and stronger ties between stakeholders and the waterbody they affect. This approach requires the three-tier analysis be completed prior to designation of critical areas since the purpose is to link sources and activities with water-quality impairments and threats. For Michigan watershed projects[3] partnerships identify the critical area and then complete an inventory of the critical area to refine the pollutant, sources, and causes. Such approaches need to incorporate an overall watershed screening to ensure pollutant hot spots are not overlooked outside the initial critical area delineation.

Not all sources of a pollutant are equally significant in causing beneficial use impairment in the receiving water. In terms of development and redevelopment-associated urbanization, urban land use with imperviousness of less than 15% is critical for protection purposes. For other nonpoint sources, designating certain areas as critical in need of management is an efficient approach to defining the extent, type, and location of land treatment needed to yield the greatest water-quality benefit. Targeting assistance to critical areas is a lesson learned for nonpoint source pollution from the Section 108(a) Black Creek Project in Indiana (1972–1977). Targeting has since been used successfully in other nonpoint source control programs, specifically the Clean Lakes, the RCWP, and the Section 319 supported efforts.

Specific criteria for targeting areas depend on the water-quality problem and watershed characteristics. Targeting criteria can be grouped into five broad categories.

Assessment and Problem Identification Phase

1. Type and severity of water resource impairment or threat
2. Type of pollutant
3. Transport considerations (distance to watercourse and to impaired water resource)
4. Pollutant source magnitude
5. Other project- and regulatory-specific criteria

The planning committee should establish targeting criteria based on the watershed assessment information generated through the three-tier analysis prior to developing objectives for land treatment.

Critical-area delineation has two components: pollutant source and pathways. A preliminary identification of critical areas needs to occur during the assessment phase. Three general types of critical areas are found in every watershed. The first type is the discrete direct discharge points to the waterbodies; the second is the area adjacent to or near the waterbody; and the third are areas that may contribute large amounts or high concentrations to the waterbodies regardless of their locations. The demanding steps of pinpointing and quantifying those sources are the first steps to effective management of critical areas. Examples of criteria used in some of the RCWP projects are proximity of animal feedlots and waste application patterns and soil erosion indices. Scientific evidence does not support using proximity to waterbodies as the sole criterion for defining critical areas.

The first step in delineating critical areas is to map source areas for pollutants of concern and relating these to threatened and impaired waters. For example, the phosphorus index is designed to obtain a relative indication of the risk of phosphorus loss in runoff on a field or site-specific basis. The index considers both transport factors and phosphorus source factors at the field level and uses a multiplicative approach to arrive at the final phosphorus index rating. In order to formulate a hydrologic basis for identification of critical areas in the watershed, site-specific analysis of the hydrologic flow paths is needed to identify the origin of surface runoff and the most contributing areas. Since dissolved constituents can be transported long distances through ditches, drain tiles, and channels, it is important to include these systems in the hydrologic analysis. Once water is channeled, the distance to a waterbody is not an obstacle to transport dissolved contaminants. To identify the most critical areas for minimizing pesticide loss to waterbodies, one must identify the hydrologic process that transports the most pesticides to waterbodies.

Critical-area designation can change based on new information gleaned during the course of a project. Critical areas in the Garvin Brook Watershed (Minnesota) were substantially redefined after 4 years of implementation using new information about groundwater problems. Monitoring data indicated groundwater, in addition to surface water, was contaminated. The project area was expanded to include the entire groundwater watershed (approximately 50% greater than the surface watershed). Additionally, the critical area was redefined to include groundwater protection. Areas classified as having very high or high sensitivity to groundwater pollution were now considered critical. This redefinition of critical areas resulted in an increase in the overall acreage needing treatment.[7]

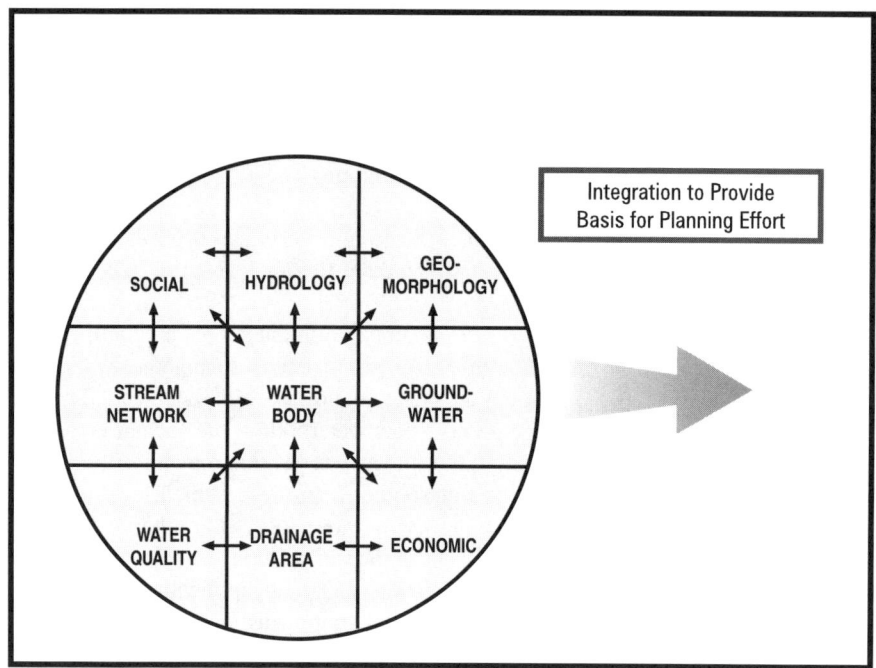

FIGURE 5.8 Integration to provide basis for planning.

Where monitoring is not possible, surrogate indexes can be used. In the Highland Silver Lake RCWP Project, implementation specialists utilized a potential pollutant index to work with landowners/operators in gaining acceptance of their role as generating nonpoint source pollution problems and designing management approaches for their contributing cropland.[25] The index was based on combining the enrichment ratio (soil-specific) and estimated total sediment loss. The index was then utilized to evaluate the relative effectiveness of various alternative management systems for the fields (parcels) under review.

5.6 PULLING IT TOGETHER

Synthesis focuses the assessment on the interaction among land-use activities, watershed processes, and resource conditions. This step is important because by identifying the causes of the pollutants' sources, the technical committee can design the most successful control and prevention program. This effort combines knowledge gained about individual components of the watershed into a comprehensive understanding (Figure 5.8).

While the watershed assessment process provides a framework for examining watershed function and status, each watershed management plan must be developed with consideration of the unique physical, hydrological, and land-use patterns that influence the state of the basin. The compilation and review of information, combined with field reconnaissance and screening-level watershed modeling utilizing assessment and comparative analysis tools, provide a basis for developing implementation approaches. Once the sources of impairment-causing pollutants are identified and quantified as to their significance, management's approach to address them at the source will be developed during the watershed-planning phase.

Table 5.1 is an example of how to organize assessment, condition, and possible source information.

At a minimum, the assessment and problem identification report should consist of:

1. A description of the project area
2. A description of the water(s) of concern
3. An assessment of water quality (where are designated uses supported and where are they impaired?) and natural resources
4. An assessment of pollutant loads
5. An identification of the critical conditions in terms of flow and season of the year during which designated uses are not supported
6. A delineation of critical areas
7. Source location and magnitude
8. An identification of areas that require additional attention, which is one of the fundamental purposes of setting and monitoring benchmarks

For more information on watershed assessment, see Dunne, T. and Leopold, L.B., *Water on Environmental Planning*, W.H. Freeman and Co., San Francisco, 1978.

Federal Interagency Stream Restoration Working Group, Stream Corridor Restoration: Principles, Processes, and Practices, www.usda.gov/stream_restoration, 1998.

For more information on urban assessment and management, see Schueler, T.R. and Holland, H.R., Eds., *The Practice of Watershed Protection*, Center for Watershed Protection, Ellicot City, MD, 2000.

For more information on agricultural assessment and management, see Ritter, W.F. and Shirmohammadi, A., Eds., *Agricultural Nonpoint Source Pollution, Watershed Management, and Hydrology*, CRC Lewis Publishers, Boca Raton, FL, 2001.

TABLE 5.1
Surface-Water Resource Conditions, Problems, and Nonpoint Source Pollution Sources in the Lower Big Rib River Watershed

Sub-Watershed	Stream Name	Length (mi)	Biological Use[1]/Miles	Habitat Rating[2]	Biotic Index Rating[3]	Problems Limiting Factors[4]	Observed or Potential Sources[5]
Marathon	Big Rib R.	10	WWSF/10	Good	*******	HB,SD,NU,F	NMM,CL,SD
	Artus Creek	7	WWFF/7	Good	Very good.	HB,SD,NU,F	URB,PSM
	Creek 3–7	0–0.75	LFF/0.75	Poor	********	HB,SD,NU	SB,SBP,CL
	Creek 5–14	0.75–1.8	LAL/1.05	*******	********	HB, SD,NU	SB,CL
		2.9	UNK/2.9	*******		HB,SD	CL
Kennedy Creek	Kennedy Creek	7	WWFF/7	Fair to poor	Very good	HB,SD,NU,F	SB,SBP,CL
Lake Wausau	Big Rib R.	4.5	WWSF/4.5	Fair	*******	HB,SD,NU,F	URB

[1] Biological use (existing and potential): This column indicates the use supported by the stream as defined in NR 102(04)(3), Wis. Adm. Code under fish and wildlife uses.

WWSF: warm-water sport fish communities; WWFF: warm-water forage fish communities; LFF: limited forage fishery; LAL: limited aquatic life; UNK: unknown.

[2] Habitat rating: This column indicates the relative quality and quantity of aquatic life habitat in the stream.

[3] Biotic index rating: The column indicates water-quality condition based on Hilsehoff Biotic Index, which uses macro invertebrates as an indicator of organic pollution.

[4] Problems limiting factors

HB: habitat (lack of cover, sedimentation scouring, etc.); SD: sedimentation (filling in of pools); NU: nutrient enrichment; DO: dissolved oxygen (low conditions); F: flooding of fluctuating water levels (flashiness).

[5] Observed or potential sources

CL: cropland erosion; SB: stream-bank erosion; SBP: stream-bank pasturing; BY: barnyard or exercise lot runoff; PSM: point source, municipal treatment plant discharge; NMM: nonmetallic mining (granite/gravel); URB: urban runoff.

REFERENCES

1. Terrene Institute, *Delineating Watersheds: A First Step towards Effective Management*, Terrene Institute, Alexandria, VA, 1993, p. 4.
2. Texas Natural Resource Conservation Commission (TNRCC), Texas Water Watch Manual for Conducting a Watershed Land Use Survey, GI-232, TNRCC, Austin, 1997.
3. Brown, E., et al., *Developing a Watershed Management Plan for Water Quality: An Introductory Guide*, Michigan Department of Environmental Quality, Lansing, 2000.
4. Illinois Environmental Protection Agency (IEPA), Guidance for Developing Watershed Implementation Plans in Illinois, draft, IEPA/BOW/98–002, IEPA, Springfield, 1998.
5. Lehre, A.K., Collins, B.D., and Dunne, T., Post-eruption sediment budget for North Fork Toutle River drainage, June 1908–June 1981, *Zeitschrift fur Geomorphologic Neue Folge Supplementband*, 46, 143–163, 1983.
6. Center for Watershed Protection (CWP), Draft Guidance on Watershed Mapping, Section G, Practical Watershed Planning for Growing Watersheds Workshop, Chicago, IL, CWP, Ellicot City, MD, 1998.
7. USEPA, Evaluation of Experimental Rural Clean Water Program, EPA-841-R-93–005, USEPA, Washington, D.C., 1993.
8. USDA-SCS, Inventory and Evaluation Procedure for Land Treatment, USDA-SCS, Champaign, IL, 1984.
9. National Research Council, *New Strategies for America's Watersheds*, National Academy Press, Washington, D.C., 1999.
10. Plafkin, J. L. et al., Rapid Bioassessment Protocols for Use in Streams and Rivers: Benthic Macroinvertebrates and Fish, EPA/444/4–89–001, USEPA, Washington, D.C., 1989.
11. Federal Interagency Stream Restoration Working Group, Stream Corridor Restoration: Principles, Processes, and Practices, www.usda.gov/stream_restoration, 1998.
12. CTIC, *Groundwater & Surface Water: Understanding the Interaction, a Guide for Watershed Partnerships*, CTIC, West Lafayette, IN, 1995, p. 16.
13. USEPA, Indian River Lagoon Septic System/Carrying Capacity Study, USEPA, Atlanta, 1996.
14. Rosgen, D.L., The use of color infrared photography for determination of sediment production, in *Proc. Fluvial Process and Sedimentation*, Canadian National Research Council, Edmonton, AB, 1973, pp. 381–402.
15. Roseboom, D.P. and White, W., The Court Creek restoration project, in *Proc. XXI Conf. Erosion Control: Technology in Transition of the International Erosion Control Assoc.*, Washington, D.C., 1990, pp. 25–40.
16. Riley, A., *In Restoring Streams in Cities, a Guide for Planners, Policymakers, and Citizens*, Island Press, Washington, D.C., 1998, p. 423.
17. Rosgen, D.L., Development of a river stability index for clean sediment TMDLs, in *American Water Resource Assoc. (AWRA) Proc. of Wildland Hydrology Meeting in Bozeman, Montana*, D.S. Olson and J.P. Potyondy, Eds., AWRA, Middleburg, 1999, pp. 26–36.
18. Schueler, T.R, The importance of imperviousness, *Watershed Restoration Techniques*, 1(3), 100–101, 1994.
19. Schueler, T.R., Crafting better urban watershed protection plans, *Watershed Protection Techniques*, 2(2), 329–337, 1996.
20. CWP, Estimating Current and Future Impervious Cover, Section H, Practical Watershed Planning for Growing Watersheds Workshop, Chicago, IL, Center for Watershed Protection, Ellicot City, MD, 1999.

21. Caraco, D.R. et al., *Rapid Watershed Planning Handbook, a Comprehensive Guide for Managing Urbanizing Watersheds*, Center for Watershed Protection, Ellicot City, MD, 1998.
22. City of Olympia, Impervious Surface Reduction Study, draft final report, City of Olympia, 1994, p. 183.
23. Winn, J., Land use planning project, *Land and Water*, Jan/Feb 1998, p. 50.
24. USEPA, Water Quality Conditions in the United States: A Profile from the 1998 National Water Quality Inventory Report to Congress, EPA-841-F-00–006, USEPA, Washington, D.C., 2000.
25. Davenport, T.E. and Kelly, M., Water Resource Data for Highland Silver Lake Monitoring and Evaluation Project Madison County, Illinois phase IV, IEPA/WPC/86–001, Illinois Environmental Protection Agency, Springfield, 1986.

Figure 5.3. Phillips, N., *A Watershed Approach to Urban Runoff; Handbook for Decision Makers,* 2nd ed., Terrene Institute, Alexandria, VA, 2002.

6 Plan Development

Plan backwards — set objectives and trace back to see how to achieve them, even though you may discover there is no way to get there and you will have to adjust the objectives. Plan forward — to see where your steps will take you. It seldom is clear and is certainly not always intuitive.

— **Donald H. Rumsfeld**

6.1 INTRODUCTION

Watershed management occurs on many scales and for a variety of purposes. The planning process provides the focus for developing a road map for addressing the goals, selecting the best management alternatives and implementation approaches, defining challenges/opportunities, and determining how to measure progress and success. The specific goals of a watershed management plan need to be defined, based on stakeholder participation, within the context of existing conditions in the watershed and taking the future into account. The goals should be establishing a desirable future condition that accommodates important economic, social, and recreational uses — and the plan provides the framework to achieve them. Watershed planning is nothing more than a systematic approach to formulating goals and objectives and laying the groundwork for accomplishing them

The process of planning is inherently long-term and outcome-oriented. Without a long-term plan, year-to-year decisions are often inconsistent and management is invariably inefficient. Effectively managing a watershed requires a systematic plan based on the best available assessment of the watershed's natural, economic, and social features. The planning process must get the science right. Otherwise, the plan might focus on an unsolvable problem based on existing technology. Plans need to focus on achieving outcomes rather than becoming a collection of tools or strategies for further analysis. Unfortunately, watershed management plans are typically reactive to problems rather than proactive to prevent future degradation. They must find a balance so the plan can be used to prevent future problems while addressing the existing problems. A small investment in pollution prevention today will prevent much greater future losses in pollution-related damages. Proper planning prevents disjointed decision making and facilitates the organization of activities. This chapter provides guidance on how to establish a plan to achieve this.

6.2 BACKGROUND

In the 1960s and 1970s local, state, and federal agencies developed comprehensive plans for regional waste-treatment plants, public land management, and watershed management. Many of these comprehensive plans for solving pollution problems were never fully implemented. These early plans shared an important characteristic; various authorities usually imposed them from the top down without participation from the people most affected by the plan. Remember, early stakeholder involvement is a key in developing watershed plans.

Why else are thousands of watershed plans sitting on shelves across the U.S.? What is wrong with them? The most common answer is the lack of funding. The Center for Watershed Protection interviewed a cross section of planners, municipal officials, consultants, and others about the wide gulf between local watershed planning and implementation. The uniform consensus was that most plans failed to protect their watersheds adequately. The survey highlighted the 11 most common reasons cited in the interviews about why local watershed management plans were not successfully implemented.[1] The results of the survey as well as other related observations are listed here.

> Reason 1: The plan was conducted at too great a scale. Scale is considered a critical factor in preparing an effective local watershed plan. The focus of plans for watersheds greater than 50 or more mi^2 was too general. DeShazo and Garrigan[2] recommend utilizing watersheds around 50 mi^2 or slightly larger when addressing a single, well-defined issue at first so that the partnership can show success. Designate the subwatershed to the scale so that the amount of land needing treatment or new management is less than 10,000 acres. In addition, the larger the watershed, the more difficult it is to ensure adequate public participation. The smaller-size watershed encourages local involvement.
>
> Reason 2: The plan was a one-time study rather than a long-term and continuous management commitment. In a number of cases the goal becomes completing the plan rather than implementing solutions. The local watershed management effort becomes a report or study rather than a process to solve problems.
>
> Reason 3: The watershed management process lacked local ownership. When consultants or a few select technical staff complete the plan, it lacks a sense of ownership from the political and public interests. According to Schueler,[1] "An over reliance on technical consultants often means that few local staff have much ownership or understanding of the plan, and consequently, have little stake in the outcome of the watershed management process."
>
> Reason 4: Plan skirted real issues about land-use regulation in the watershed. Some plans take a static view of land use and fail to provide a framework for addressing cumulative impacts of expected land-use changes. In a number of cases, agriculture was not adequately addressed because of political pressure and lack of understanding.

Plan Development

Reason 5: Budget for watershed plan was low or unrealistic. Watershed management requires the commitment of financial and staff resources over a substantial period. Most watershed plans fail to incorporate the true cost of implementation, i.e., outreach, monitoring, and O&M.

Reason 6: Plan focused on the tools of watershed analysis rather than their outcomes. In a number of situations the plan becomes a demonstration of technology and tools rather than a plan to address specific water quality. GIS, modeling, and mapping with remote sensing are some of most common tools and technologies employed in this manner; I discuss them later.

Reason 7: Document was too long or complex. While watershed management needs to be based on good science and public involvement, these factors do not have to be covered in detail in the watershed management plan. The plan needs to focus on what will be done, who is responsible, when and where it will be done, how it will be tracked and evaluated, and what the estimated cost will be. Decision makers frequently could not even find the specific watershed management recommendations they were to support to ensure implementation.[1]

Reason 8: The plan failed to critically assess the adequacy of existing local programs. Most watershed management expands on existing programs and efforts as a foundation for the new efforts without evaluating their present effectiveness. In many situations the watershed effort is needed because the existing efforts do not adequately address water-quality concerns. Agricultural areas have tremendous political pressure to use the existing voluntary approach with additional financial incentives and technical assistance rather than assessing the existing voluntary effort, identifying barriers, and developing an approach to address the issue of faulty farm management or if farming is an appropriate land use within that watershed.

Reason 9: Plan recommendations were too general. This reason is directly related to the issue of size. Recommendations need to be specific in terms of what needs to be done, who will do it, how much it will cost, and where and when (expected completion time) it will be done.

Reason 10: Plan has no regulatory meaning. This means authority to compel the implementation or use of the plan in day-to-day decision making within the watershed is nonexistent.

Reason 11: Key stakeholders are not involved in developing the management plan. Stakeholders come together under the premise that they can share information and resources, overcome conflict, identify priorities, and accomplish common objectives. Stakeholders must be involved from the beginning of the watershed management process, particularly so in plan development. The purpose of using this process is to allow stakeholders a legitimate and early opportunity to participate in the plan's development. Stakeholder involvement and ownership facilitate obtaining public reaction, building consensus, and developing support for implementation.

Dr. Peter Nowak[3] identified the most important reason plans fail technically, known as the "Lake Wobegon" factor. Dr. Nowak points out that by focusing on aggregated averages, planners lose critical information about finer-scale biophysical, social, and institutional processes. Planners need to forget the average and the appeal to equality and target the few vulnerable watershed locations that cause the majority of the problems.

Watersheds are not only hydrological units with administrative boundaries, they consist of communities. While not listed as a reason, the lack of political will or backbone to support implementation is another major reason local watershed plans failed to meaningfully prevent or reduce cumulative impacts at the watershed level over a sustained period of time. The lack of political support is usually related to (1) funding and the politicians' inability to dedicate funding because of already high local taxes, or (2) some recommendations being unacceptable to affected stakeholders. Throughout the nation, watershed restoration efforts have been successful where locally led planning processes have been used to focus political force toward finding solutions to local problems. The affected community is essential to planning and implementing management practices. This was the case for Otter Lake, Illinois, where the local water commission viewed elevated levels of atrazine as a community problem that needed to be corrected with its leadership. The Otter Lake Resource Planning Committee (OLRPC) was thus formed; it included farmers, ag-chemical dealers, seven communities, conservationists, and Water Commission representatives. The OLRPC headed a locally led planning effort to create a long-term implementation plan to reduce atrazine levels in the lake. After the plan was implemented, atrazine levels in drinking water dropped below three parts per billion (the drinking-water standard), and the watershed's soil, animal, and plant resources improved.[4]

Born and Genskow[5] conclude the following about watershed plans and planning processes:

- Watershed plans should get people involved.
- Watershed plans should be condensed so that the public easily understands them.
- Watershed plans are never "done" — each version sets the stage for the next version.
- Watershed plans need a long-term monitoring component.

6.3 PLANNING

Strategic planning is rarely done solely as an intellectual activity. The exercise is used to focus decision making, shape partnership policies, motivate stakeholders, and guide a host of other activities. The first rule of strategic planning is to make sure the right objectives have been developed. The second rule is to periodically revise the plan to incorporate new information through an iterative approach to maintain balance between area-wide problems and subwatershed-specific problems. The strategic planning effort needs to use a logical progression from screening, targeting, and ranking critical areas to management plans and include a description of how implementation activities will be tracked and evaluated. The management strategy must take into account staff availability, size of the project area, extent of critical area, and landowners' receptivity to the selected management practices.

Plan Development

Every watershed partnership needs to adopt a planning process that is dynamic and flexible to meet changing conditions and needs, is structured to organize information and decision points, and provides for the iterative use of tools to maximize use of existing tools and minimize development costs. High levels of participation are expected in seeking out, debating, and synthesizing knowledge related to important assessment issues. Partnerships must demonstrate commitment through action. The watershed plan does not need to be completed before implementation activities can begin. The first and most critical action for a partnership, where possible, is to control pollutant sources causing degradation or limiting water use. Using the plan development process, a partnership can mobilize resources to address clearly identified problems now.

The four aspects of watershed planning — the type of plan, the purpose of the plan, the unit of government sponsoring the plan, and the planning method employed — are critical to determining the actual plan. Written watershed management plans are supporting, not prescriptive, tools. They must be flexible. No plan is static. Things change, additional information and data are obtained, and people's needs, values, and attitudes change. Plans have to change systematically to accommodate the changes. However, plans that are modified too quickly to changing demographics and political whims fail to provide the necessary stability that successful long-term management requires. A plan that is too rigid, though, will be useless and ignored in short order.

The goal of the assessment and identification phase is to provide the information necessary to identify the specific nature and cause of impairments or threats throughout the watershed and to support informed decision making in selecting and designing management alternatives. The information must be scientifically valid and easily presented to and understood by the public. A fundamental rule of watershed planning is to set proper public expectations for the effort. One effective way is to acknowledge uncertainty, promote evaluation of performance, and have a process to make adjustments as part of the management effort from the start.

The changes between the existing conditions and future projected conditions, due to the implementation of an alternative, are the effects of that alternative. Watershed management planning involves a set of choices; positives and negatives are associated with these choices. Evaluating alternatives is the process of determining which alternatives are the most effective, efficient, complete, and acceptable in solving problems and meeting the stakeholders' objectives. The selected alternatives may not optimize all four aspects. The output from this analysis becomes input for the planning phase. In correcting the causes of problems, the danger always remains that treating one symptom of impairment could trigger another unwanted change in conditions.

Three decisions must be documented in the watershed management plan to maintain stakeholders' confidence in the effort's achieving its natural resource and environmental goals. The plan must explain how the actions were put in priority order, how implementation will be documented, and how the effort's achievements will be measured. The plan will have documented three determinations:

1. The project's results (overall natural resources and water-quality goals)
2. What actions are needed to achieve the results (match solutions to particular problems for the watershed and identify the implementation process)
3. What actions can and should be taken

6.3.1 PLANNING PHASE

During this phase the planning and technical advisory committees:

- Identify and refine management opportunities. Look for cheaper and low-maintenance opportunities, consider various levels of pollution controls, and increasingly refine and reevaluate solutions
- Accurately define the decision points within each step of the phase. Iteratively reevaluate priorities, consider public input and other considerations, consider cost constraints at each iteration, and consider compliance requirements
- Verify management opportunities. Define additional data needs
- Identify additional studies. Define needs for more detailed analysis

The committees need to consider multiple constraints when developing a watershed plan:

Financial constraints
Economic constraints
Development and population growth
Technical and scientific limitations
Regulatory and compliance constraints
Ownership constraints

These constraints provide the social and economic limits to the planning effort. Unfortunately, many of these constraints are rarely discussed explicitly, much less keeping in mind implicit tradeoffs. The ownership constraint provides a tremendous opportunity for the partnership when it comes to public land management. The public ownership constraint can be a benefit if the federal, state, and tribal governments agree to support the watershed management effort by making the management of their proprietary lands consistent, to the maximum extent practicable, with the watershed management plan. The age-old efficiency/effectiveness and short-term/long-term dilemmas often lie at the root of these tradeoffs. The partnership needs to develop water-quality management options, within constraints, based on an early and inclusive mix of public and private interests. When only a few interests are represented, the perception is that something is being done that is not quite right. The sooner everyone is involved and informed, the better.

During this phase, various alternative management options and approaches are publicly examined and eventually reconciled into a plan that works for all stakeholders. Otherwise, some stakeholders may not agree with parts of the plan and feel alienated from the decision-making process. These stakeholders frequently stop participating and resist plan implementation.

Two sets of goals, long and short term, and corresponding objectives and actions to achieve each goal must be established. The planning team should use a three-tier list — goal, objectives, and actions — in developing alternative management approaches. The partnership must be realistic in developing clear short-term mile-

Plan Development

stones and long-term goals. (Clear goals must be set so that they are easy to explain.) Also important is developing clear, concise, focused goals and objectives to guide the management effort, no matter how small or large, and ensuring all those involved agree about the goals and objectives. It is easier to get support, e.g., financial support and volunteers, if stakeholders and the public understand what the partnership is trying to do, and how it relates to them. Short-term milestones are often the most difficult to set and are the most important because they relate directly to the resource addressed. Short-term milestones are also necessary to ensure the project remains focused. The milestones must be measurable or quantifiable. Finally, these milestones should relate to pollutants in the watershed and to the targeted water uses. Without clearly stated objectives, the partnership will later find it difficult to select and endorse solutions to the identified problems.

The partnership needs to establish a set of targets and then not waiver from those targets; if necessary, change activities to ensure meeting targets. The partnership must know what it wants to achieve and commit to achieving it. An example of the use of interim milestones in an adaptive management process is the Lower Big Rib Priority Watershed Project. The evaluation plan for the 10-yr project consisted of an annual administrative review, pollution reduction evaluation, water resource monitoring, and final report. The pollution reduction goal was 3907 tons of sediment from critical cropland areas, which meant targeting approximately 117 fields owned by 65 landowners. The plan had an interim milestone related to addressing the primary pollutant sediment originating from the critical cropland areas. The interim milestone was adopted; after 5 years, the amount of sediment projected to be reduced was to be equal to or greater than 2344 tons measured against cost-share agreements. Failure to meet this interim milestone would result in adding approximately 426 new fields to the critical-area category.[6]

Using both interim and final targets is particularly well suited to watershed management efforts in which it might take many years to attain final targets and water-quality standards because of the slow response of the waterbodies to land-use changes. Analysts and stakeholders want clearer short-term measures to guide near-term implementation and evaluate management effectiveness, if the analytical basis for final target levels is weak.

The link between the initiative and the response and then the subsequent adjustments create the dynamic that determines the success of the implementation effort. The long-term incremental nature of watershed work makes it essential that the partnership finds ways to determine if it is heading in the right direction and if it is making progress fast enough. Benchmarks to measure progress fall into three general categories:

1. Organizational benchmarks focus on monitoring the partnership and how it works.
2. Activity benchmarks focus on documenting efforts expended to improve water quality.
3. Watershed benchmarks are environmental measures related to the management effort.

The problems associated with impaired waters took many years to reach the current state. It is unrealistic to believe that they will be solved in a very short timeframe. Develop plans and resource projections to match this long-term commitment to water-quality improvement and protection.

6.3.2 PRIORITY SETTING

In linking specific land management activities and watershed health, the planning committee must establish specific measurable conservation targets linked to waterbody goals. Conservation targets are quantifiable objectives that can be used to prioritize land-based activities and to evaluate progress against water-quality improvements as surrogate tracking measures.

Prior to proceeding to the development of specific objectives, the planning committee needs to work with the overall organization to determine the committee's priority goals. Time and funds are never adequate to address all potential watershed management needs. Priorities must be established that target efforts to the most critical problems and opportunities. The planning committee needs to work with the steering committee on prioritizing actions by establishing criteria. Common criteria in priority-setting processes include the ability to influence changes, delay between actions and results, willingness to change, and cost–benefit ratio. Extensive participation from the partners and stakeholders in a debate on the partnership's key strategic issues is encouraged. Three key strategic issues to gather stakeholder input are the availability of financial and human resources, their time constraints, and water-quality issues. Prioritize critical water uses in order of importance for stakeholders.

No single, best method for prioritizing pollutants exists. It can be based on (1) the priority the partnership gives each pollutant for each designated use, (2) the number of designated uses the pollutant impairs or threatens, and (3) what the partnership feels it can accomplish first. It is important to select and document a process acceptable to both steering committee and stakeholders. For the fictitious Ethan Watershed, the technical committee developed pollutant rankings for each designated use and then used these rankings to develop an overall ranking of pollutants. The committee ranked sediment as number one because (1) it impacts two designated uses, and (2) nutrients are often attached to sediment. Addressing sediment also reduced nutrients. Criteria commonly used for ranking pollutant sources included the frequency of occurrence, the degree to which the source degraded the water, and an analysis of benefits.

Creating focus, setting priorities, and shaping implementation approaches are related to partnership effectiveness. The second set of attributes focuses more on the underlying needs of the stakeholders. Prioritization is a combination of art and science: knowing what is most important to accomplish and understanding what is, in fact, accomplishable, given the partnership's resources. This process needs to consider:

- Implications of management actions on prioritization of problems
- Implications of management actions on prioritization of subwatersheds
- Implication of anticipated future conditions
- Solutions that reduce technical or construction complexities

TABLE 6.1
Cost Effectiveness by Management Practice, in Terms of Phosphorus Reduction

	Watershed 1		Watershed 2	
Management Practice	Phosphorus Reduction (lb)	Effect $ per lb of Phosphorus Reduced	Phosphorus Reduction (lb)	Effect $ per lb of Phosphorus Reduced
Barnyard runoff control	686	9	171	31
Milkhouse wash treatment	75	26	24	71
Waste storage	340	593	31	4328

Because resources are limited, the planning committee must prioritize management efforts and management practices to those subwatersheds that provide the highest return on resource investment in terms of meeting the overall watershed goals. Although many different methods exist to set priorities, the specific method chosen is not as important as having an open and well-understood process for establishing priorities. Meals[7] demonstrates the importance of this with an evaluation of several manure management practices on the phosphorus export from two subwatersheds in the LaPlatte River Basin in Vermont. Both subwatersheds exhibited a lower phosphorus export after management practice implementation, and barnyard runoff control was the most cost-effective management practice. Table 6.1 shows the phosphorus reduction and effectiveness by management practice and subwatershed. In a theoretical application of this concept, using the information in Table 6.1, the watershed planning committee, with limited resources, should select subwatershed 1 as its highest priority. Also, because the cost-effectiveness of individual management practices varies considerably, the planning committee should emphasize barnyard runoff controls. An evaluation of benefits, or benefit maximization, is another way to determine cost-effectiveness of management practices. Benefits traditionally fall into three categories:[8]

1. Prioritized benefits, which are ranked by preference or priority.
2. Quantifiable benefits can be counted but not priced. If benefits can be quantifiable on a common scale, a cost per unit of benefits that identifies the activities that most efficiently produce benefits can be devised.
3. Nonmonetary benefits can be described in monetary terms.

A key in evaluating benefits is the impact of the activity in terms of scale and value and its timing. The short- and long-term benefits of the management approaches must be estimated.

Watershed management plans to protect or restore water resources can be developed to correspond to support programs, policies, and goals already established within the community, such as those outlined in comprehensive plans, economic development plans, transportation plans, and recreation plans. The Ythan project is an excellent example. Recently, a project to protect, improve, and enhance the Ythan River in Aberdeenshire, Scotland, was approved for funding under the European LIFE Environment fund. The Ythan project is the first large-scale designated nitrate vulnerable zone in Scotland. The Scottish EPA identified fertilizers as one of the main source of nitrates that seep into groundwater or run off into the river network. Using information developed by the Scottish EPA, the Formartine Partnership, which includes representatives of local and national organizations, coordinated the planning and funding request. Formartine Partnership Chairperson Neville Jones said, "The work will include helping farmers to apply for government aid to support more environmentally friendly practices. Farmers will also get help with completing nutrient budgets for their farms to ensure fewer nutrients are lost into the river, we'll also be restoring some specific sections of the river." The project's goal is to take in other crucial aspects of the quality of life of the river and local community, including fishing, environmental education, woodland management, and wildlife enhancement. By focusing on all these issues, the partnership aims to help make the watershed more attractive to tourists and people who live and work there.[9]

6.3.3 WATERSHED STRATEGY DEVELOPMENT

The water-quality and natural resource problems and data gaps identified during the assessment phase set the stage for the plan development phase. The identified water-quality problems are evaluated in the context of the interdependency of the geomorphological and ecological features and social needs in order to develop a watershed management plan. A key to the planning process is for the partnership to view storm runoff not as a nuisance but as a resource; the partnership needs to work with the hydrologic cycle rather than against it. Rainwater needs to be viewed as an urban asset. Phillips[10] promotes a watershed management strategy development that follows this hierarchal list:

Protect what you have.
Prevent further degradation.
Restore natural sediment and water regime.
Restore natural channel geometry.
Restore riparian plant community.

A key to this hierarchy is controlling the basin's hydrology and hydraulics to maintain the health and stability of downstream waterbodies and their tributaries.

Figure 6.1 depicts the increased intensity and complexity along the management continuum.

The first four steps of Phillips' hierarchy can be related to water-quality management, for which two basic approaches to implementing pollution controls exist: the water-quality approach and the technology approach. The water-quality approach requires the development of a plan to control pollutants exceeding the water-quality standards. Recently this has also been referred to as the TMDL approach. The technology-based approach requires current water pollution treatment technology installed whether pollutants discharge into clean or impaired waters. An example of the technology approach for municipal wastewater discharge is secondary treatment; for agricul-

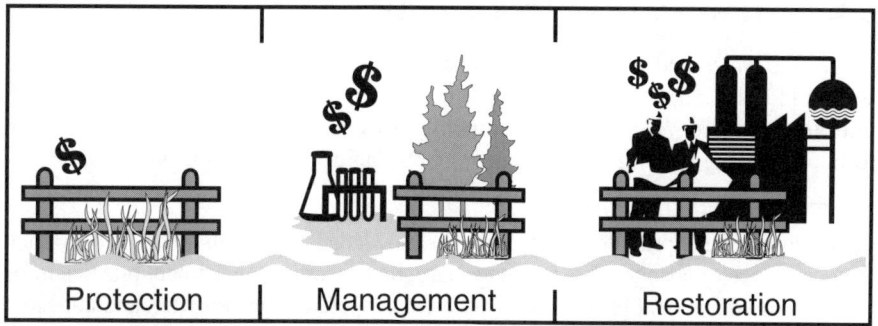

FIGURE 6.1 The scale of costs per activity. (Reprinted with permission from Terrene Institute, Alexandria, VA, as published in *A Watershed Approach to Urban Runoff; Handbook for Decision Makers,* 2nd ed., 2002.)

tural row crop production it is a CORE4 system. CORE4 is a common sense approach to improving farm profitability while addressing environmental concerns. CORE4 integrates residue management, crop nutrient management, weed and pest management, and conservation buffers to sites specifically to help farmers manage their lands.

Pollution control measures for both point and nonpoint source pollution benefit the watershed but often may not provide an economic return to the individual, industry, or municipality that has to install them to meet watershed management plan goals. Selecting controls may thus be a point of contention; planning members may argue for the least costly measures, while others may prefer the most effective controls, regardless of cost. A common myth is that equitable tradeoffs can be made between point and nonpoint source reductions and in the long term the environment will be maintained and protected. For the same pollutant, the relative impact of one lb of pollutant from a point source vs. one lb of pollutant from a nonpoint source discharger has never been scientifically documented. The ratios of nonpoint source-to-point source reductions are a result of the imprecision of the ability to estimate nonpoint source loading rather than seasonal or long-term impacts associated with various types of pollution sources. The easiest way to address this is for the planning committee to adopt performance-level standards for all pollutant sources and leave the specific pollution control designs to the implementation team.

Adopted management approaches must control both permanent and temporary sources of pollutants. The traditional focus of watershed efforts has been on the control of permanent increases in pollutant export caused by irreversible changes to the landscape. However, equally significant are temporary increases in pollutant export caused during the disturbance phase in land-use conversion — this must also be addressed.

Current stakeholders must continue to maintain existing management practice systems and pollution controls even if the economics of the situation change. The planning committee must identify the action items necessary to maintain existing practices if functioning properly and, if not, work with the owners to ensure proper maintenance and operation. For newcomers, the cost of pollution control needs to be explained as a part of doing business in the watershed.

6.4 IMPLEMENTATION APPROACHES

Part of the watershed management plan should be an explanation of how key issues were addressed and what the partnership will do to get the public more involved after the plan is developed. Four distinct approaches are used to overcome barriers to implementing the watershed management plan activities and tasks.

Information and Education

Outreach activities provide the awareness, knowledge, and skills needed to promote the occurrence of change. This facilitates targeting specific management practice information and educational programs to landowners/operators. Information and education activities, part of an overall social capacity building component, explain the financial and technical assistance available in a watershed and promote awareness of what needs to be done for the problems and issues addressed. One-on-one contact with landowners and farm operators in the critical area can help speed up implementation.

Technical Assistance

Making contacts and providing technical assistance to a large number of landowners is time-consuming. Plan for sufficient staff to support the level of effort required to achieve objectives. Regulatory responsibilities and requirements are explained through educational efforts. Watershed management efforts rely heavily on social capacity building to be successful. Behavior changes identified in a watershed management plan take time to get in place. Watershed management planning efforts need to realize this and ensure an adequate social capacity building effort is in place because information needs to be given over and over again to be effective. Information and crucial messages need to be presented in different ways to match the different target audiences' learning methods. A strong foundation of the water problems and goals needs to be conveyed, so when stakeholders are asked to change behavior they know why and are likely to support doing it.

Funding or Financial Support

Direct cost-share assistance, indirect subsidies through price supports or taxation systems, and subsidized technical support help overcome the economic barrier. While technical assistance is normally not considered financial assistance, it is included as a type of financial assistance since it is provided free of charge to landowners, operators, and businesses within the watershed. The value of this assistance needs to be accounted for and conveyed to the landowners.

Regulatory Options

Federal, state, and local regulations, pending or in effect, help overcome the barrier of inertia. Regulations raise the priority for people to make a behavioral change. However, regulations usually prevent people from doing wrong things, but they provide no incentive for doing the right things. It takes an extensive outreach effort to explain the regulation and what it takes to comply with it in order to build public compliance.

Ownership of the effort is covered in its long-term success. Stakeholders need to give their time and effort; they provide examples for change when they see the watershed management effort as important to them, their family, and their community. Stakeholders need to feel ownership of projects to ensure project support.

Plan Development

The implementation effort has three primary components: outreach, management approach, and monitoring and reporting. Each component has a unique goal under which the activities are grouped. The goal of the outreach component is to inform and educate the public about the overall project and provide specific information on selected topics to targeted audiences. The goal of the management approach is to apply sufficient pollution control and restoration management practice on the land to improve water quality. Finally, the goal of monitoring and reporting is to track and evaluate the progress of all project activities.

6.4.1 Agriculture

Six basic watershed approaches support voluntary nonpoint source controls. These six approaches are built on two concepts: stewardship and targeting. The four main combinations of these concepts are stewardship, targeting, landscape, and any combination of the previous three. The stewardship and landscape approaches allow all landowners in the watershed to be eligible for some type of assistance. This helps build a broad base of support from an effort, while simultaneously diffusing efforts so that the partnership may not be able to document an impact on off-site resources such as waterbodies. The targeted and combination approaches result in the partnership providing assistance to only eligible landowners. The stewardship approach distributes technical and financial resources equally rather than in a prioritized manner. The stewardship approach is relatively ineffective because the technical and financial resources are not targeted to the main pollutants of concern and their primary sources. In contrast, the water-quality-based approach requires identification of the pollutant causing the water-quality impairment and then directs focus of the land management improvements to critical pollutant source areas. This approach effectively addresses the pollutant source and transport to impaired (or threatened) water resources. Unfortunately, the stewardship approaches are the most prevalent and the least effective in addressing water-quality impairments. Dr. Nowak[3] provides several reasons for the institutional support for the stewardship approach:

- Institutional reward systems are structured such that field people are penalized for focusing on disproportionate contributions.
- Prevailing technology transfer models are based on trickle-down models, thereby penalizing other approaches.
- A social data infrastructure is lacking.

Targeting programs raises the issue of equity in rural areas where the typical USDA broad-based stewardship programs have been implemented. The targeted approach focuses on just addressing critical pollutant generation and pathway areas rather than the stewardship approach, which promotes the proper management of all lands by all people.

The basic approaches are listed below.

1. The "first-come, first-served" approach is where assistance is provided to all audiences in the watershed. Figure 6.2a shows the management boundary for this approach.

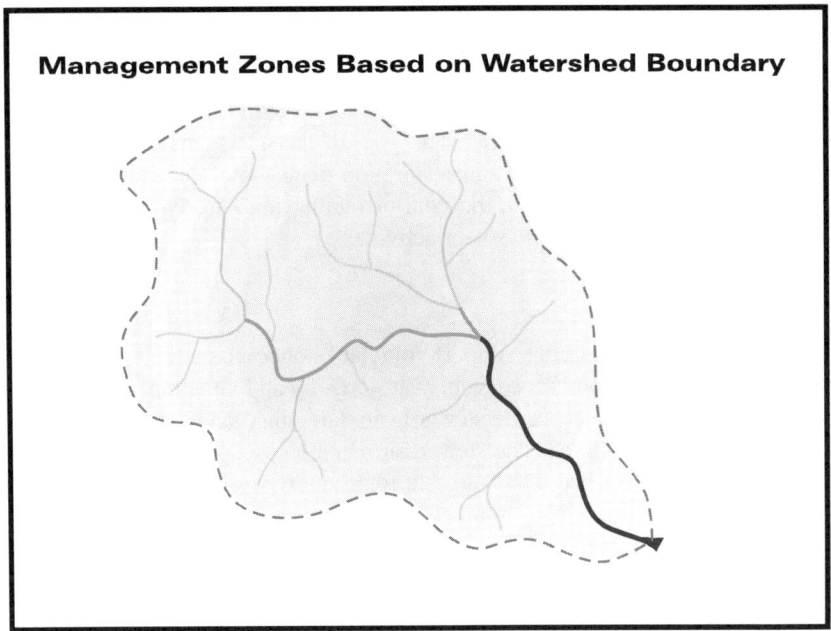

FIGURE 6.2A Management zones based on watershed boundary. (Courtesy of CTIC, West Lafayette, IN.)

2. "Cooperator" assistance is offered to receptive audiences, which are usually the landowners who have cooperated with a past or ongoing effort.
3. The "trickle-down" approach focuses the implementation efforts on community leaders who, in turn, influence early adopters, who influence their neighbors.
4. The "critical-site" approach requires implementation to focus on those areas where biophysical criteria indicate high vulnerability to contributing to off-site degradation. The goal in determining critical areas is to match resource needs with targeted efforts to get the greatest benefits. A high level of treatment in critical areas will result in a greater reduction of pollutant delivery as compared to a broad-based approach with lower levels of land treatment over larger areas. These areas depend on the watershed and the watershed goals. Figure 6.2b shows this critical-site approach using proximity to the waterbody as the biophysical criterion.
5. The "worst-first" approach is where the audiences targeted are those "critical sites" where landowners exhibit inappropriate behavior.
6. The "landscape" approach is where types of assistance are targeted to various zones and critical areas within the watershed. The goal in determining critical areas is to match resource needs with targeted efforts to obtain the greatest benefits. The delineation of these areas depends on the watershed and the watershed goals. Figure 6.2c shows an application of the management zone approach based on sources and transport potential.

Plan Development

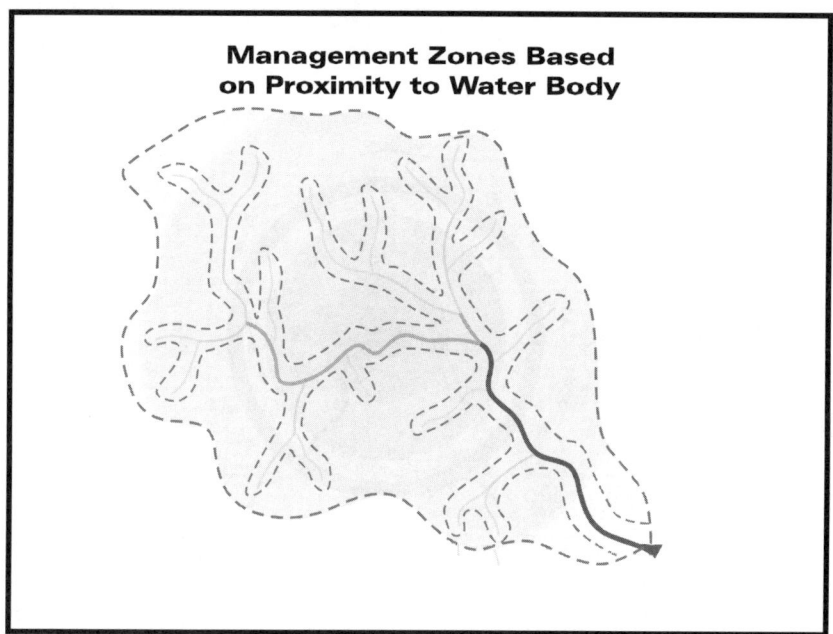

FIGURE 6.2B Management zones based on proximity to waterbody. (Courtesy of CTIC, West Lafayette, IN.)

For watersheds where critical areas have not been identified, every landowner is eligible for all available assistance or where the steering committee wants to offer something to all land operators/owners. One way to enhance the effectiveness of a stewardship effort is to use a landscape zone approach for applying pollution controls and determining eligibility for different types of assistance on a watershed basis.[11] The watershed's landscape can be divided into general zones, based on a combination of spatial juxtaposition and dynamic interaction between potential pollutant sources and runoff processes and the resulting water quality, and then managed accordingly.

The most efficient approach is to divide the watershed landscape into three zones: upland, transition, and riparian. The upland zone is characterized as relatively flat, less than 2% slope, with permeable, well-drained soils and the absence of sheet wash erosion. The relative magnitude of the eroding force of sheet wash is less than resistance of the soil to sheet wash, so sediment production would be minimal in this zone. Rain-splash erosion does occur in the upland zone. The transition zone has slopes greater than 2% and slopes toward a waterbody or collection area. There are usually ephermal conveyance systems in the transition zone; depending on the land use and management, erosion can be extremely high (50 tons/acre). The riparian zone is adjacent to a waterbody, relatively flat with both depositional and conveyance areas. For dissolved pollutants, the drainage (natural and manmade) network becomes the overriding factor rather than spatial location of the pollutant generation areas. The partnership must determine the management level needed to adequately control pollutant loadings. The landscape approach lends itself to establishing vari-

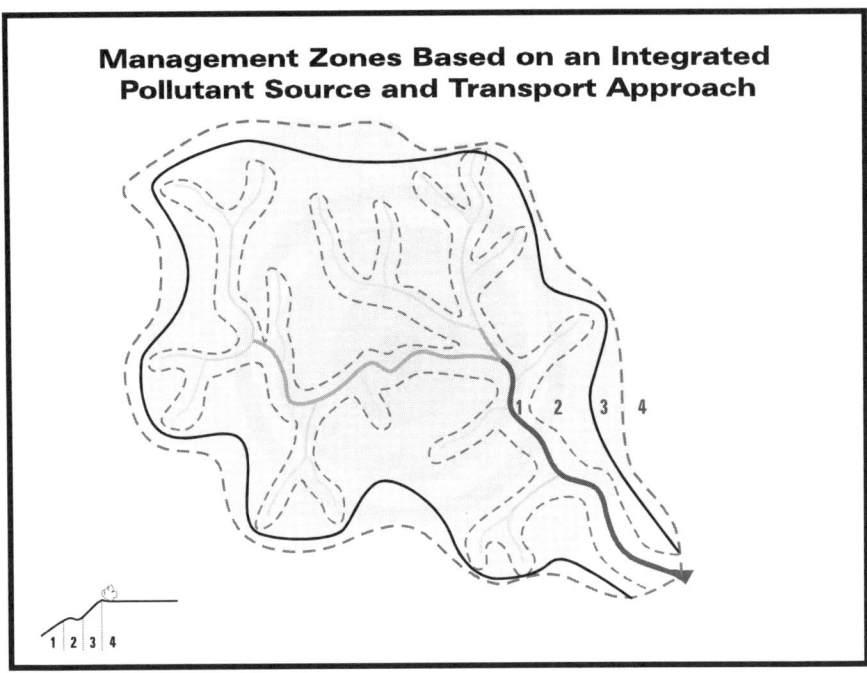

FIGURE 6.2C Management zones based on an integrated pollutant source and transport approach. (Courtesy of CTIC, West Lafayette, IN.)

able management objectives based on the pollutant of concern and in which zone the area is located. A management goal for upland areas would be pollutant source reduction. For the transitional zone the goals would be source reduction and pollutant pathway management, and for the riparian zone it would include pollutant treatment in addition to source reduction and pathway management.

Assistance can be targeted to each landscape zone. Availability of technical assistance is a large part of this component. The first step is to identify which agricultural service sector has expertise in the practices being promoted. The second is to identify market benefit the agricultural service industry can realize by promoting the management practice. For example, landowners/operators in all three zones (riparian, transitional, and upland) would receive outreach assistance whereas a landowner whose operation is in the riparian zone and contributes pollutants to the adjacent waterbody would be eligible for both financial and technical assistance. Transitional zone landowners would be eligible for technical assistance. Those in the upland areas would not be eligible for these types of assistance unless a site-specific determination proves it was contributing a significant load to the waterbody. To incorporate the "worst-first" concept, an initial screening approach is recommended for the transitional and upland zones in order to identify any critical areas for pollution generation or needs for a more detailed assessment. The screening level is designed to isolate the potential problem spots based on location, severity, and potential consequence or landowner behavior.

Plan Development

6.4.2 Urban

Management of water resources is an extension of government's police power, included under the government's power to protect public health, safety, and general welfare. The state delegates this power to local units of governments through legislation. Communities need to identify whose responsibility it is to protect water quality and get the public involved, and base delineations and plans on sound science. For urban areas, the watershed plan sets the goals and objectives on a watershed basis, identifies the basic watershed management strategies to meet the goals and objectives, and provides specific directions to guide land-use planning decisions through the development of subwatershed plans. The water resource requirements on individual land parcels through site-specific plans such as a stormwater management plan and development conditions that will meet the goals and objectives of the subwatershed plan are usually governed at the municipal level. The specific management techniques or management practices are implemented through site plan controls, stormwater management plans, subdivision agreements, and erosion and sediment control guidelines or standards. These components constitute levels of planning but also aspects of watershed management when implemented. In the municipal land-use planning process, the key planning document is the official plan. At all levels, clear responsibilities are assignable to appropriate agencies.

A community embarking on the development of a watershed management effort should further refine its goals to include specific, measurable, and achievable objectives. These objectives include measures such as:

- Minimize impervious areas to preserve groundwater recharge and drinking-water source protection areas
- Provide an equitable sharing of costs and benefits of protecting water quality
- Protect water supplies from adverse effects of urbanization
- Preserve open-space land for aesthetics and recreation while also preserving land for water quality

For the watershed approach to work, municipalities must have a critical role in the development, implementation, and evaluation of the management effort. Although boundaries of watersheds rarely align with political jurisdiction, municipalities bring together three essential components for management:

1. Authority to regulate how land is used
2. Authority to construct and manage wastewater treatment facilities
3. Authority to protect and treat drinking water

Municipalities can provide leadership within a watershed partnership by initiating efforts to:

- Partially restore the predevelopment hydrological regime by reducing the frequency of bankfull flows in the contributing watershed

- Reduce polluted stormwater pulses by retrofitting ponds and wetlands, watershed pollution prevention program, and the elimination of illicit connections to sewer networks
- Stabilize channel morphology and restore equilibrium channel geometry
- Restore in-stream habitat structure
- Reestablish riparian cover as an essential component of the structure ecosystem
- Protect critical stream substrate to support fish spawning and secondary protection by aquatic insects
- Allow for recolonization of the stream community by removing fish barriers and selective stocking of native fish to recolonize the stream reach

Good wastewater treatment is critical to successful long-term watershed management. Regardless of the treatment system, all require proper design, operation, and maintenance. These requirements vary among treatment types. Overall, it is most efficient and cost-effective to collect wastewater from homes and industries and treat it in one large facility than to have individual septic systems or treatment systems or facilities. Typical waste treatment systems for large cities and municipalities are conventional sewer systems piped to treatment facilities. These facilities include components such as activated sludge, biofilters, contact stabilization, land treatment, and large-scale lagoons. Most municipal treatment systems have both primary and secondary treatments. Some treatment systems include tertiary treatment (see www.epa.gov/owm for more detailed information).

Stormwater programs under the NPDES Phase II Stormwater Programs must include the following six minimum control measures: public education and outreach; public participation/involvement; illicit discharge detection and elimination; construction site runoff control; postconstruction runoff control; and pollution prevention/good housekeeping.[12] Planning and zoning plans developed in response to the NPDES stormwater requirements are predicated on the use of watershed hydrologic models that can evaluate the impacts of existing and proposed stormwater management practices and management options on a watershed. Many options are available to manage existing sources of contamination and to ensure that future land-use activities do not pose a threat to water quality. Urban watershed management must integrate wet-weather (combined sewer overflows and stormwater) issues with upstream drinking-water source protection efforts. This concept takes full advantage of the multiple pollution reduction — political and environmental goals to foster watershed initiatives that enhance the health of the watershed's waterways and the public's perception of the environment.

Philadelphia is employing this approach to improve water quality and enhance the city's waterways.[13] Their initial efforts, led by the city's Public Works Department (PWD), are focused on the Darby and Cobbs Creek watershed. The PWD is promoting an approach to water-quality management that seeks to reduce water pollution from all sources based on measurable results — increased dissolved oxygen and decreased fecal coliform levels in the streams, stream-bank restoration, and the addition of riparian buffers to adjoining park land, or a mixture of both. This effort is geared to reconnect the city with its waterways, to make the streams and parks valuable community assets that will induce citizens to join in watershed management efforts. In response to the public education mandate, a number of homeowner guides focusing on pollution prevention/good housekeeping have been developed. Connecticut's *TIPS for Clean Water* is an excellent example.[14]

The lack of a common theme has often made it difficult to achieve a consistent result at the individual parcel level or cumulatively, at the watershed scale. One of the best strategies a municipality can employ is minimizing the aggregate amount of new impervious surfaces. Watershed-based zoning is based on the premise that impervious cover is a superior measure to gauge the impacts of growth, compared to population density, dwelling units, or other factors. Once such scheme is outlined below.[1] It divides urban streams into three management categories based on the general relationships between impervious cover and stream quality:

1. Stressed streams (1 to 10% impervious cover)
2. Impacted streams (11 to 25% impervious cover)
3. Degraded streams (26 to 100% impervious cover)

The level of the imperviousness in a subwatershed and the uses of the water resource determine the level of management required and the potential to allow new or redevelopment in the subwatershed. The key steps are assessment, measuring imperviousness, designating the future stream quality, and adopting specific resource objectives for stream and subwatershed. To address runoff quantity issues as a means to reduce downstream flooding and pollutant transport, on-site detention needs to be promoted. On-site detention can be accomplished through the implementation of site design measures that limit impervious surface area, buffer strips, practices that maintain natural drainage, wetlands, reforestation, rain barrels, and rainwater cisterns. Lehner and colleagues[15] report that the Staten Island Blue Belt project uses natural systems and processes to control flooding and prevent stormwater pollution in a relatively undeveloped part of New York City. Working with stakeholders, the city developed a plan that emphasizes using existing natural drainage features as part of a stormwater system servicing 6000 acres. Settling ponds, constructed wetlands, and sand filters provide water-quality and quantity control in areas with existing storm sewers. The projected savings for the city exceeded $50,000,000 by avoiding construction costs associated with a subsurface sewer system. These savings included the cost of land acquisition. In addition to the water-related benefits, the community preserved open space and natural habitat.

Imperviousness is one of the variables that can be explicitly quantified, managed, and controlled with a watershed. Build-out analysis is a component in discussions concerning future conditions of the resources within a watershed. Build-out analysis typically uses existing land-use zoning and master plan requirements to project the future land use and density of the use for a particular area. Population forecasting is another technique commonly used to estimate growth patterns and its impact. Build-out analysis (using zoning maps) is commonly used to estimate impervious area. Neither of these techniques is completely accurate. Communities seldom reach maximum possible build-out, so using the zoning map approach generally overestimates future impervious coverage. Population forecasting is usually inaccurate

TABLE 6.2
Estimated Loading Rates for Buttermilk Bay

Source	Concentration (mg nitrogen/liter)	Nitrogen Loading Rate	Flow/Recharge
Wastewater	40	6.72 lb/person-yr	55 gal/person-day
Fertilizer (lawns)	—	0.9 lb/1000 ft^2-yr	18 in./yr
Fertilizer (bogs)	—	15.8 lb/acre	—
Pavement runoff	2.0	0.42 lb/1000 ft^2-yr	40 in./yr
Roof runoff	0.75	0.15 lb/1000 ft^2-yr	40 in./yr
Acid precipitation	0.3	3.03 lb/acre	—
Dwelling	—	25.3 lb/yr	—

Source: Phillips, N.J., *Decision Making for Watershed Management Using Build-Out Analysis*, Conservation Technology Information Center, West Lafayette, IN, 1999.

because it reflects current growth patterns that may or may not be sustained. Build-out analysis offers great flexibility in considering many types of alternatives. Alternate build-out analysis is constructed from the existing condition incorporating various management approaches and resource considerations. Physical management alternatives for a number of natural resources such as wetlands, sole source aquifers, critical slopes, riparian zones, stream corridors, buffer requirements, and mitigation corridors can be included. In combination with predicative models, a build-out analysis can be used to project water quality and quantity changes as a result of a projected change in land use. In this context, predicative models allow planners to examine various management practice scenarios for the alternatives to estimate the relative range in net change in water quality and quantity. To aid in the decision-making process, the existing condition is then compared with the various alternatives to estimate what further degradation may be incurred by selecting one alternative versus another. This is particularly important when developing TMDLs that incorporate future growth. An excellent example of using build-out analysis to evaluate management approaches is the Buttermilk Bay analysis presented in Table 6.2.

Municipalities have many options in regulating and managing land use. Good land-use planning assists in watershed management by avoiding problems with which to begin. Health regulations, zoning ordinances, land acquisition, and voluntary controls are some of the options local government has in its efforts to protect and manage water resources.

Phillips[16] demonstrates the feasibility of utilizing build-out analysis in managing nitrogen loadings from a coastal watershed in Massachusetts. Buttermilk Bay is a 214.5-hectare (530-acre) shallow coastal embayment located at the northern end of Buzzards Bay (Figure 6.3). Portions of three towns (Bourne, Plymouth, and Wareham) are located in the watershed. Algae blooms, elevated chlorophyll concentrations, and declining eelgrass beds in localized areas are the documented symptoms caused by excessive nitrogen loading from stormwater runoff. Based on this information a four-step study was initiated. First the surface and groundwater drainage areas were delineated. Second, existing and potential levels of development, accord-

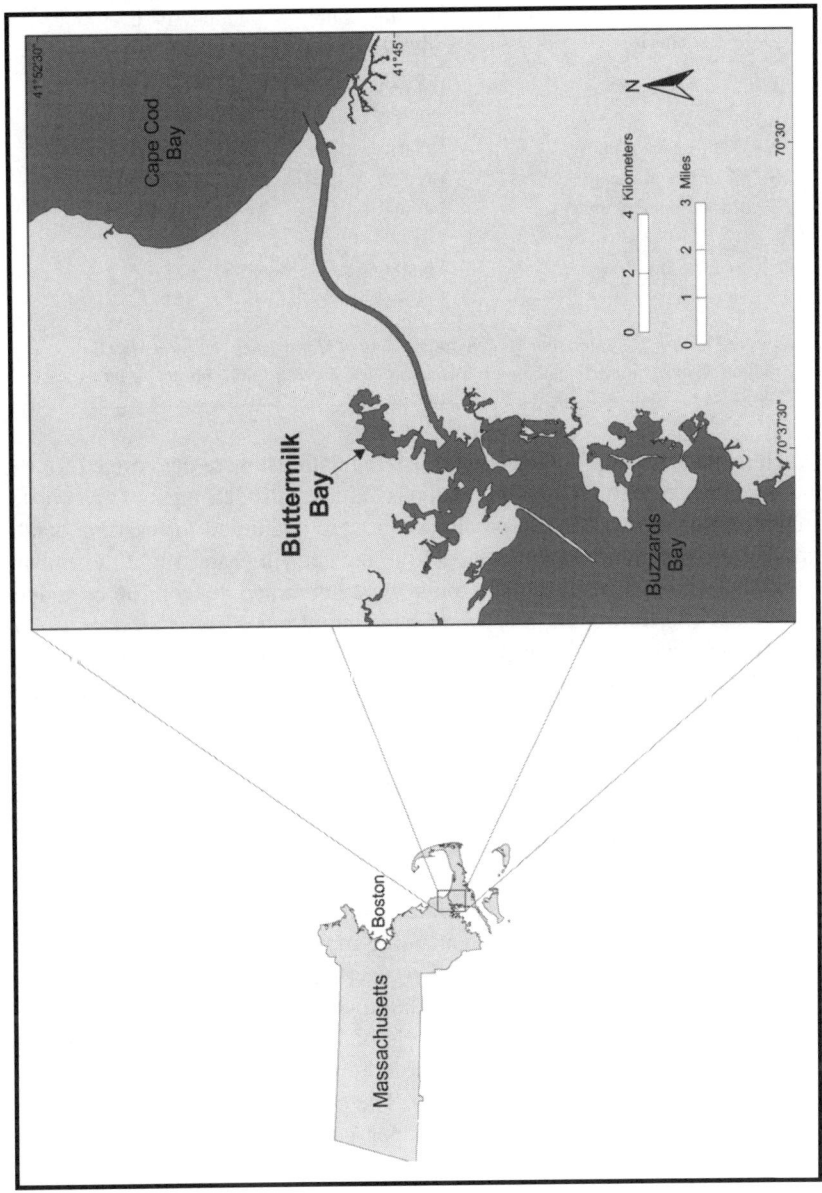

FIGURE 6.3 Buttermilk Bay, Massachusetts. (Courtesy of CTIC, West Lafayette, IN.)

TABLE 6.3
Nitrogen Loading by Alternative Scenario

Scenario	Annual Loading Rate (lb/yr)	Nitrogen Loading in Buttermilk Bay (mg/m^3/R)
Existing condition	91,053	189
1	126,664	263
2A (3% of lots at 4 units)	133,044	276
2B (7% of lots at 4 units)	141,550	294
2C (10% of lots at 4 units)	147,980	307
3	110,185	229
4	157,486	327
5	123,662	257

Source: Phillips, N.J., *Decision Making for Watershed Management Using Build-Out Analysis,* Conservation Technology Information Center, West Lafayette, IN, 1999.

ing to each respective town's current land-use regulations, were determined for the contributing area through a build-out analysis. Third, nitrogen inputs from wastewater, fertilizers, road runoff, and precipitation were quantified for existing conditions and build-out conditions and compared to the carrying capacity of the embayment. Fourth, land-use management tools were identified to control or reduce nitrogen inputs. The Buttermilk Bay Watershed's build-out analysis used a baseline condition of 3049 residential and 39 commercial units and indicated the potential for an additional 2265 residential units under full build-out conditions for all developable land.

A critical loading rate of 240-milligram/cubic meter/residence time (mg/m^3/R) was recommended for Buttermilk Bay. Seven alternative build-out scenarios were examined under various conditions (Table 6.3). The scenarios varied in development intensity, management conditions, and natural resource restrictions.

Scenario 1: An additional 2265 residential units would be constructed in the watershed with an average occupancy rate of 3 people per unit. This scenario incorporates the existing sewer projects.

Scenario 2: Assumes a full build-out condition with an occupancy of 3 people per unit and a percentage of the parcels of land within the watershed of more than 20 acres are subject to a comprehensive permit for affordable housing at densities of 4 units per acre. The percentage of parcels of more than 20 acres with a density of 4 units per acre varies with 3, 7, and 10% of the lots.

Scenario 3: Assumes a full build-out condition with an occupancy rate of 3 people per unit and the utilization of small wastewater treatment plants on parcels of land greater than 30 acres in size. Tertiary treatment with an effluent nitrogen concentration of 10 mg/L is assumed.

Plan Development

Scenario 4: Assumes full build-out condition with an occupancy of 3 people per unit, full commercial development for nonresidential parcels, and full utilization of agricultural parcels for agricultural production.

Scenario 5: Assumes full build-out condition with an occupancy of 3 people per unit, with 5% of currently undeveloped land reserved as permanent open space distributed evenly across the watershed.

Table 6.3 shows one scenario is compatible with the critical loading rate of 240 mg/m^3/R for Buttermilk Bay. Scenario 3 assumes a full build-out condition of an additional 2265 units with an average occupancy of 3 people per unit and utilizing small-scale wastewater treatment plants on parcels of 30 acres and greater in size. Tertiary treatment is the prescribed level of management with an effluent limit of 10 mg/L. However, scenario 1 is a more realistic prediction of future land use for the watershed. The predicted nitrogen load for scenario 1 is 126,664 lb/yr, which represents an excess of 11,187 lb/yr, or the equivalent of approximately 442 single-family dwellings. Adjusting either lot size or adding additional management practices to scenario 1 could result in an acceptable nitrogen load. For example, increasing the minimum lot size to 70,000 ft^2 would allow only 1823 new dwellings and restrict future loading to approximately the critical level for the bay.

6.4.3 Urban Riparian and Wetland Management

Wetlands are water resources of the U.S. and need the same degree of attention and protection as other waterbody types. Research shows that watershed hydrologic stability is a factor of the wetland-to-watershed ratio. Results of Illinois State Water Survey studies indicate that in order to maintain hydrologic stability the wetland-to-watershed ratio should be at least 12%.[17] The Conservation Technology Information Center (CTIC)[18] identifies seven ways a watershed can benefit with wetlands:

1. Improve water quality by breaking down, removing, using, or retaining nutrients, organic waste, and sediment carried with runoff from the watershed.
2. Reduce severity of floods downstream by retaining water and releasing it during drier periods.
3. Protect stream banks and shorelines from erosion.
4. Recharge groundwater, potentially reducing water shortages during dry spells.
5. Provide food and other products — such as commercial fish and shellfish — for human use.
6. Provide fish and wildlife, including numerous rare and endangered species, food habitat, breeding grounds, and resting areas.
7. Increase opportunities for recreation — bird watching, waterfowl hunting, photography — and outdoor education.

Watershed management efforts should include a wetlands component with a goal of attaining at least 12% wetlands-to-watershed ratio and include an evaluation of the existing wetlands resources: identifying areas where wetlands need to be protected from nonpoint source pollution; areas with the potential to enhance existing degraded wetlands for water-quality purposes; and areas for consideration of wetlands restoration.

The watershed plan should clearly document the various priorities by water resource, including wetlands. In terms of ecological priority, wetland restoration is substantially preferable to wetland enhancement (making a wetland "better") or wetland creation (building a wetland where there never was a wetland). For existing wetlands preference should be given to protecting water-quality functions and having a system that is self-maintaining. This level of performance would be expected for restored or enhanced wetlands. Community education and targeting protection/management techniques, such as zoning, to wetlands and natural areas that focus on the water-quality benefits and value of wetlands can be part of the watershed management plan.

Urban riparian zones are giving some rivers room to breathe and the community a valuable resource. Riparian zones are green spaces, ideally extending at least 100 ft along a watercourse and containing a succession of habitats that filter pollutants and reduce erosion. Riparian zones also help maintain hydrologic stability through storage and extended discharge of floodwaters as well as the interception and transpiration of precipitation. Riparian zones with a functional value are those on natural floodplains or low natural areas adjacent to perennial waterbodies. Riparian zones in the upper portions of the watershed are more valuable for hydrologic control than those lower in the basin. Restored and enhanced wetlands can be a major component of any headwater protection and riparian zone management program. They provide everyday services such as filtering polluted runoff, controlling floods, and providing habitat for wildlife. These services have been valued at $5973/acre.[19] While the first priority is to protect existing wetlands, restored and constructed wetlands can provide the same benefits as natural wetlands.

A major issue with wetlands and riparian zones is land ownership. Private interest's providing a service for the overall community without financial compensation is a concern for landowners who believe they have been denied full utilization of their land. Selection of sites should consider landowner/community willingness; the cost of land, restorability, maintenance, and landscape position are also issues. Public or NGO land acquisition has become a valuable tool in watershed management.

Barnegat Bay, New Jersey, is a classic example of integrating coastal development and natural resource conservation. The Barnegat Bay watershed is located in one of the fastest-growing counties in the U.S. Its year-round population almost doubles during the peak summer vacation season. The primary threat to both the Barnegat Bay ecosystem and the drinking-water supply is nonpoint source pollution associated with development. Water treatment alone is an impractical solution to the nonpoint source pollution since the approach does not address the ecosystem impacts. In 1987 the New Jersey Legislature ordered a study of the environmental threats to the Bay's watershed. A watershed plan was developed that called for the creation of buffer zones through the acquisition of sensitive areas. The Trust for Public Lands[20] published *The Century Plan: A Study of One Hundred Conservation Sites in the Barnegat Bay Watershed*. The plan identified 100 high-priority conservation and public access sites in need of protection. A 1997 follow-up study identified

Plan Development

other vulnerable lands. The 1997 study prioritized the top 10 areas based on the following five criteria:

1. Importance to water quality
2. Importance as wildlife habitat
3. Level of disturbance with preference to undisturbed properties
4. Adjacency, or proximity to already protected properties
5. Size, with a preference for properties large enough to offer significant benefit

So far, land acquisition in the Barnegat Bay watershed, funded through a variety of programs, totals approximately 120,000 acres. A local trust fund was established. Other funding sources include the Land and Water Conservation fund and the Migratory Bird Conservation Fund for wetlands and adjoining uplands. Sixty-one percent of Ocean County residents supported a ballot measure to create a local Open Space Trust Fund, funded through a new property tax, to match available funding from the New Jersey Green Acres program. This indicates a high level of stakeholder support. The social capacity building of the stakeholders and developing the initiative in the context of an overall watershed effort are key to the success of a program such as this.

6.5 RESTORATION OPPORTUNITY

Restoration opportunity is determined by the circumstances and timing of what needs to be accomplished. The two primary economic reasons why restoration can be more cost-effective than advanced point source controls are that (1) restoration often has lower marginal costs, and (2) restoration provides a wider range of ecological benefits. For example, lake restoration projects must be combined with watershed management activities in order for a lake choked with sediment to be transformed from a public nuisance into a viable pubic resource. Even with watershed management activities addressing existing problems, restoration efforts must anticipate and be able to respond to future upstream disturbances such as land-use conversions caused by urbanization and deforestation. Restoration is not solely applicable to severely degraded environments. Although it can be used as an effective tool to return a degraded system to a predisturbance condition, restoration is also an important tool for preventing further environmental degradation. Strengthening structural and functional elements through restoration can help improve a system's tolerance to stressors that could lead to environmental degradation.

Three options are available when restoration is needed:

1. Nonintervention and undisturbed recovery are applicable where the waterbody of concern is recovering rapidly and active management is not necessary.
2. Partial intervention for assisted recovery is applicable where a waterbody is attempting to recover but is doing so slowly or uncertainly. In these situations, action is needed to facilitate natural processes already occurring.

3. Substantial intervention for a managed recovery is applicable where a waterbody has desired functions that are beyond the repair capacity of natural processes and active restoration measures are needed.

The goal of any of these three options is to establish self-sustaining waterbody functions while still accommodating economic, social, and recreational uses.

Intuitively, the goal of restoration should be first to replace what has been lost. If the causes of the impairment can realistically be eliminated, complete watershed restoration to a natural or desired condition might be a feasible objective and a long-term focus of the watershed management plan. When the causes of impairment cannot be eliminated, it is critical to identify what options exist to manage either the causes or symptoms of the impaired conditions, and what effect the various management approaches might have on the waterbody. While mitigating the impacts of the disturbances is an alternative method, it is not preferred. When mitigation is chosen, the focus of the effort is then on addressing only the symptoms of the impairments rather than reducing or eliminating the impairments. Cairns[21] identifies several sources of difficulty in implementing restoration efforts: multiple interest groups and objectives; lack of measurable criteria; myriad large and small decisions under uncertainty; and interdisciplinary cooperation in public setting. Proper planning can minimize the difficulties Cairns identified.

For restoration components, the implementation team must define a reference or desired condition, establish a baseline condition, develop and implement a post-construction evaluation, and use an adaptive management approach. The adaptive management component provides for continual, scientifically informed fine-tuning of the restoration activities and associated upstream management. In a climate of limited financial and technical resources, it is often necessary to geographically prioritize restoration efforts based on overall watershed benefits. Under this paradigm, resources are then allocated to geographic areas where the functional benefits from restoration are the greatest.

A hierarchy exists in approaching in-system restoration with upland or surrounding watershed techniques; the riparian techniques are followed as needed by in-stream techniques. Aquatic restoration needs to be a composite of in-stream, riparian, and upland techniques. Balancing and integrating in-system, riparian, and surrounding watershed approaches are essential. Any restoration plan could involve a combination of techniques, depending on environmental conditions and stressors to be addressed.

When considering restoration components in the planning process, it is important to use a systems approach. Everything in a systems approach should emulate natural systems. Then the partnership can build an economically superior system. Restoration activities for purposes other than pollution control must be scheduled for implementation or installation after the causes of the problem have been addressed. After arresting degradation, the plan must provide the framework to accelerate the recovery of habitats impacted by destabilization of the channels and downstream waterbodies. This includes using structures to create pool habitat, planting to reestablish riparian vegetation, and stabilizing actively eroding banks with bioengineering techniques. Then a series of structures is used to create a pool/riffle habitat, reestablishing riparian vegetation, and modification to develop a new floodplain

Plan Development

within the incised channel or reconnecting of the stream network to its original floodplain is completed. The Waukegan River (Illinois) project documented the applicability of this approach.[22] This situation needed to go beyond just pollutant control and install in-stream restoration practices to improve water quality; stream stabilization was supplemented with the reestablishment of pool/riffle complexes to improve dissolved oxygen levels and increase habitat. In other situations installing management practices to restore function and structure aspects is sufficient. The Skokie River restoration project had three main goals: to stabilize the eroding channel banks, to restore the riparian buffer zones beyond each bank, and to improve water quality and habitat within the stream itself. The physical effort was to make the Skokie River narrower and deeper, thus making it cooler. Using a combination of techniques including brush layering, willow posts, and coir logs, some sinuosity was returned to the river.[23] Follow-up site inspections indicated that the project is achieving its goals. A number of lake watershed projects have shown that in-lake measures are necessary to actualize the benefits of the watershed land treatment efforts in terms of in-lake water quality. Lake Le-Aqua-Na and the Johnson Sauk trail are excellent examples of following watershed improvements with in-lake management restoration techniques.[24,25]

Watersheds contain a wide variety of places to locate constructed wetlands for water-quality benefits. The implementation team needs to evaluate such locations in the watershed. Watershed landscapes have natural patterns that maximize the value and function of individual wetlands. Mitsch and Gosselink[26] identify four landscape locations for constructed wetlands: in-stream, riparian, upstream, and terraced wetlands. Wetlands can be designed as in-stream systems with the addition of control structures to the streams themselves or by impounding a tributary of the stream. In-stream wetlands should be considered in lower-order streams. Riparian wetlands are designed to capture sediment and associated pollutants to reduce pollutant loading during flood stages and from surface runoff. For upstream wetland construction, watershed coordination, rather than piecemeal parcel or subwatershed planning, is needed to maintain the hydrologic regime and provide the maximum water-quality benefits. Terraced wetlands are wetlands integrated into the landscape's steeper terrain using terraces to place the wetlands on the hillside.

One of the least discussed and most difficult parts of restoration efforts is maintenance. Plans must include a complete maintenance schedule that outlines roles, responsibility, and frequency. It is often assumed that maintenance takes care of itself, and delegating responsibility to local units of governments without additional support is unworkable.

6.6 CONCLUSION

Good planning takes time (Table 6.4) and has results:

1. It builds focus and direction. It explains the reason for the organization. Who is part of it? It also identifies the common ground and lays out what the organization wants to accomplish.

TABLE 6.4
Typical Management Plan Development Schedule

Activity	Timeframe (Months from Initiation of Watershed Effort)	Product/Outcome
Inventory/study	6	Data and information compiled
Assessment	8	Synthesis of information
Problem definition	10	Information combined with knowledge and values
Prioritization	11	Priority concerns and areas identified
Alternatives analysis	12	Approach selected
Plan development	18	Management plan

2. It sets the structure. It defines the decision-making authority and the respective authorities' responsibilities. It also outlines the working groups and their relationships.
3. It documents realistic assessment of resources. Financial, human, and time resources are defined for each action.
4. It provides the road map for the future. All future activities come from the plan. The vision of the future channels activities and resources.
5. It provides a process for course corrections. The plan cannot be static: it needs to change as new members are added and as exogenous and endogenous factors change.
6. It provides the foundation for measuring the success of the management efforts. The plan becomes a major evaluation tool.

Table 6.4 provides a typical schedule for the development of a watershed management plan.

The planning process does not proceed in just one direction and on a single track; several activities may go on simultaneously. New information may cause the group to revisit a decision. It is important for the partnership to be able to answer the following questions:

Does this approach address our primary concerns?
Will it help us meet our goals?
What would be the effects on the environment?
Can we measure the effects?
How long would it take to see results?
What would this approach cost?
Is the expertise available?
Is this approach acceptable to stakeholders?

The Lake McCarrons restoration project highlights the importance of two critical steps in the planning process. Lake McCarrons, an urban lake in Minnesota, had a

> **BOX 6.1**
> **Parts of a Watershed Plan**
>
> *Goals*: Describe the future condition and provide direction for long- and short-term actions. Key words in a goal statement are "to" and "for." The ultimate products are solutions and an executable plan. Public input and cost constraints have been considered at each decision point.
>
> Examples: To protect the water quality in Lake Emily for increased fishery production.
>
> To increase the amount of open space for community recreation in Marionville.
>
> *Objectives*: Objectives are specific, concrete, and measurable statements of intent. Objectives need to include what is to be done, how it will be measured, and when it will be accomplished. Key words in an objective statement are "to" and "by."
>
> Examples: To organize and implement two Lake Emily beach clean-up days during 2002.
>
> To stabilize four miles of critical stream-bank erosion along the lower third of Ethan River in 2002.
>
> *Action steps*: Specific work activities necessary to accomplish each objective. Action steps must include who will do it, what specifically will be done, and by when.
>
> Examples: The Restored Ethan River Committee will select five sites for clean-up by December 9, 2002.
>
> The education committee will set the training schedule by April 26, 2002.
>
> *Monitoring and evaluation*: Criteria must be able to distinguish between failures of science and failures of poor application of the science.
>
> Example: To monitor upstream and downstream to evaluate the effectiveness of the clean-up sites on the Ethan River.
>
> *Budget*: Hint: It might be helpful to organize resource allocations along the same lines as source assessment and linkages and document them.
>
> Examples: It cost $6000 for personnel, $2000 for equipment and supplies, and $1000 for materials to clean up the Ethan River sites.
>
> It costs $1800 to clean a site, and so the Ethan River Clean Up will be $9000.

restoration project implemented to reduce algae blooms in the lake. Ten years later, lake residents complained of persistent algae blooms and obnoxious plants. A study found that lake phosphorus concentrations had not declined, algae levels and water clarity had not improved, and nuisance aquatic plants were still a problem. While the lake had been monitored for the 10 years, progress had not been evaluated. First, the community had not been involved in establishing the direction of the project. If the community had been, it would have recognized the concerns with the nuisance aquatic plants. Second, after 10 years of monitoring, it was clear the lake's condition had not improved. No critical evaluation was built into the planning process, so the plan was never modified to better address the project's goals.[27]

The plan will answer a few simple questions: what, who, where, when, and how can the watershed manage to protect or restore? The plan needs to serve several roles. It serves as a symbolic function in that it represents the common vision of the partnership. The plan establishes a framework for addressing critical water-quality and natural resource issues, problems, and needs. The plan serves as a record of all subsequent activities by outlining the process. The plan also serves to communicate the elements of the process to the public and stakeholders. Watershed management plans should be viewed as a starting point and not the end product.

Additional resource materials include the following:

Caraco, D. et al., *Rapid Watershed Planning Handbook*, 1998, a comprehensive guide for managing urbanizing watersheds.

Holdren, C., Jones, W., and Taggart, J., *Managing Lakes and Reservoirs*, N. Am. Lake Manage. Soc. and Terrene Inst., Madison, WI, 2001.

Federal Interagency Stream Restoration Working Group (FISRWG), Stream corridor restoration: principles, processes, and practices, 1998.

REFERENCES

1. Schueler, T.R., Crafting better urban watershed protection plans, *Watershed Protection Techniques*, 2(2), 329–337, 1996.
2. DeShazo, R.P. and Garrigan, P., Merrimack River Initiative, *Watershed Connections, Lessons Learned in Subwatersheds of the Merrimack River Watershed: The Nashua, Souhegan and Stony Brook Watersheds*, New England Interstate Water Pollution Control Commission, Wilmington, 1996, p. 35.
3. Nowak, P., Dr., Interaction of Social and Bio-physical Parameters: What Needs to Be Done? Presentation at ASIWPCA/EPA Nonpoint Source Meeting, November 28, New Orleans, 2001.
4. Farnsworth, R. et al., The Otter Lake story: anatomy of a successful locally led planning effort, *Grassroots Planning*, 1, 15, 1998.
5. Born, S. and Genskow, K.D., Exploring the Watershed Approach: Critical Dimensions of State-Local Partnerships, the Four Corners Watershed Innovators Initiative, final report, River Network, Washington, D.C., 1999, p. 54.
6. Wisconsin Department of Natural Resources (WiDNR), Wisconsin Department of Agriculture, Trade, and Consumer Protection, Dane County Land Conservation Department and Marathon County Land Conservation Department, Nonpoint Source Control Plan for the Lower Big Rib River Priority Watershed Project, WT-539-00, WiDNR, Madison, 2000, p. 143.
7. Meals, D.W., LaPlatte River Watershed Water Quality Monitoring and Analysis Program: Final Report, Program Report #12, Vermont Water Resource Center, Univ. of Vermont, Burlington, 1990.
8. USEPA, Ecological Restoration: A Tool to Manage Stream Quality, EPA 841-F-95-007, USEPA, Washington, D.C., 1995.
9. Scottish Environmental Protection Agency (SEPA), Cash boost for the river Ythan, *SEPA View*, Winter 2001, SEPA, Stirling, Scotland, 2001.
10. Phillips, N., *A Watershed Approach to Urban Runoff; Handbook for Decision Makers*, 2nd ed., Terrene Institute, Alexandria, VA, 2002.

11. Davenport, T.E. and Kirschner, L.T., Landscape approach to sediment control, *7th Federal Interagency Sedimentation Conf. Proc.*, Reno, 2001, pp. 73–80.
12. USEPA, Stormwater Phase II Proposed Rule — Small MS4 Stormwater Program Overview, Fact Sheet 2.0, EPA 833-F-99–002, USEPA, Washington, D.C., 1999.
13. Dahme, J. and Smullen, J.T., Innovative strategy helps Philadelphia manage combined sewer overflows, *StormWater*, 1(1), 52–57, Nov/Dec 2000.
14. Connecticut State Museum of Natural History, *Tips for Clean Water*, Connecticut State Museum of Natural History, Hartford, 2000, p. 4.
15. Lehner, P.H. et al., *Stormwater Strategies: Community Responses to Runoff Pollution*, Natural Resources Defense Council, New York, 1999.
16. Phillips, N.J., *Decision Making for Watershed Management Using Build-Out Analysis*, Conservation Technology Information Center, West Lafayette, IN, 1999, p. 21.
17. The Wetlands Initiative (TWI), ADID Wetland No. 55 Protection Plan and Watershed Development Guidance Manual, TWI, Chicago, 1997.
18. CTIC, *Wetlands: A Key Link in Watershed Management, a Guide for Watershed Partnerships*, CTIC, West Lafayette, IN, 1995, p. 6.
19. Coast Alliance, *Pointless Pollution: Preventing Polluted Runoff and Protecting America's Coasts*, Coast Alliance, Washington, D.C., 1999, p. 42.
20. Trust for Public Land, *The Century Plan: A Study of One Hundred Conservation Sites in the Barnegat Bay Watershed*, The Trust for Public Lands, Washington, D.C., 1995.
21. National Research Council (NRC), *Restoration of Aquatic Ecosystems*, National Academy Press, Washington, D.C., 1995.
22. Davenport, T.E. et al., National nonpoint source monitoring program: documenting water quality improvements from best management practices through long-term monitoring projects, in *Proc. of TMDL Science Issues Conf*, St. Louis, MO, March 4–7, 2001, Water Environment Federation, Washington, D.C.
23. Kleine, A., A river runs thought it, *The American Gardner*, March/April 1997, 26–29.
24. Kothandaraman, V. and Evans, R.L., Diagnostic-Feasibility Study of Johnson Sauk Trail Lake, Illinois State Water Survey (ISWS) Contract Report 312, ISWS, Urbana, 1983, p. 126.
25. Davenport, T.E. and Kaynor, S., Watershed management works: the Lake Le-Aqua-Na project, *Land and Water, The Magazine of Natural Resource Management and Restoration*, 42(2), 25–27, 1998.
26. Mitsch, W.J. and Gosselink, J.G., *Wetlands*, 2nd ed., Wan Nostrand Reinhold, New York, 1993.
27. Holdren, C., Jones, W., and Taggart, J., *Managing Lakes and Reservoirs*, N. Am. Lake Manage. Soc. and Terrene Inst., Madison, 2001, p. 382.

7 The Watershed Management Plan

There is nothing more difficult to take in hand, more perilous to conduct, or more uncertain in its success, than to take the lead in the introduction of a new order of things.

— Niccolo Machiavelli

7.1 INTRODUCTION

When the planning process is completed, what does the partnership have? It has a plan, written on paper. The most important aspect of the completed plan is how it is used, not that it exists. Plans not followed are useless and a waste of time and resources. The plan is an instrument for defining what must be done, how it will be done, and who will do it. As a communication vehicle, the plan tells funders what will be accomplished and also serves as a public outreach device.

7.2 PLAN ELEMENTS

A watershed management plan needs to be an honest appraisal of the potential implementation activities needed for the management of the watershed, not a political document. The document must include all parts of the management approach. The plan's readability level should be for a person with limited knowledge about the watershed. Stakeholders should be able to read the plan and understand the watershed needs and proposed solutions for effectively managing and restoring water quality and critical natural resources. The watershed management plan is a document that:

- Defines the watershed. Identifies the name, size, administrative boundaries, and geographic location.
- Describes the partnership. Identifies public and private partners who participated in the development of the watershed management plan and who will be participating in its implementation.
- Characterizes the watershed. Describes the chemical, biological, hydrologic, and other physical characteristics; existing land uses and impervious cover; existing point sources; and applicable water-quality standards and related designated uses.

- Identifies problems, indicators, and measures. Defines critical areas related to not attaining water-quality standards; identifies deviations from water-quality standards; defines sources or source categories of contaminants; identifies causes of nonattainment by segment; identifies data and information gaps; identifies indicators including habitat conditions, species of interest, and other natural resources.
- Identifies goals and objectives. This section describes the project's general goals, the specific water-quality goals and objectives, and the implementation goals. The goal is a concise statement describing the intent of the project. Objectives for the watershed management process must be measurable and related to the goal. A measurable objective has four parts: an audience (clearly stated), behavior (what action is expected from the audience), condition (the condition needed in order for the audience to carry out the desired behavior), and degree (criteria for determining if the objective has been met). These four components result in the objective's clearly stating what is to be accomplished and what standard to use in judging the accomplishment. Objectives can be written at various level of complexity as long as they contain the components.
- Summarizes pollutant loads identified in the assessment and identification phase.
- Identifies target pollutant reductions, critical areas, priority ranking of critical areas, approach that will be utilized to manage the problems.
- Describes the implementation approach. Provides a description of the management approach that includes incentives, outreach, and specific regulatory mechanisms that will be utilized to implement management practices. The institutional arrangements employed by the involved agencies and organizations at the local level need to be documented.
- Lists the management practices' availability. This section involves the identification, description, and evaluation of management practices available for controlling and preventing pollution (point and nonpoint source) and for remediating the impacts of the pollution of waters of concern.
- Lists what permit actions are needed. This section covers the necessary construction and operation permits for completing the project and what modifications will be needed for existing water discharge permits.
- Presents milestone schedule. This is a management schedule, with short- and long-term goals and milestones, and should be provided on a relative basis (January to December) given the uncertainty of the actual starting date of implementation activities at the time of plan development. The schedule should list the tasks that will be completed within each program activity in sufficient detail to provide a road map to completion of the overall effort. To ensure that sequence of the project's activities is adequate, a critical pathways analysis for the overall watershed management plan is needed. A Gantt chart (a bar graph timeline) of the primary tasks in each program element is a useful approach to critical pathway analysis. This section must also include a description of the interim, measurable milestones for determining whether management practices are being implemented.

- Describes the monitoring and evaluation effort. An evaluation and monitoring component is required to document and evaluate the implementation of control and restoration techniques implemented and the resulting change in water quality. This section should describe the evaluation procedures that the partnership will use to identify, record, and compile the pertinent information for each implementation activity. The specific questions the monitoring component is designed to answer should be identified. The intent of the monitoring is to document changes in the quality and quantity of waters of concern; however, given the environmental constraints and variability present, an implementation project may not be able to provide conclusive evidence of water-quality changes. Describes the process by which the partnership will link monitoring data to milestones to determine if a modification to the management approach is necessary to reach the outlined water quality goals and objectives.
- Describes the outreach component. The outreach component that will be used to enhance public and target audience's understanding of the watershed management effort and encourage participation in selecting, designing, implementing, and maintaining management practices. This is an integral part of all watershed management partnerships. The outreach component needs to have sufficient, clear details about what, when, and how the activities will be completed.
- Provides management practice with operation and maintenance (O&M) processes and requirements. The long-term effectiveness of the project depends on the continual functioning of management practices installed through the partnership. Partnerships historically have ignored the importance of human and social knowledge needed to ensure the continual operation of the implemented management practices. The lack of adequate O&M procedures in past projects has demonstrated the need for the O&M component.
- Describes the roles and responsibilities. Describes the roles and responsibilities of the various individuals, agencies, and organizations to be involved in the implementation of activities. The roles and responsibilities of the partners should be summarized in enough detail to document the commitment for accomplishing the project. It is important to link implementation actions with specific roles and responsibilities. Delineation of roles and responsibilities is important to ensure that all people targeted for involvement are aware of the tasks expected of them throughout the watershed management effort. A master schedule approach can be utilized to delineate roles and responsibilities by activity.
- Provides a budget for implementing the plan. Costs are a critical consideration in watershed management. Canfield[1] said, "The most important elements in lake management are not phosphorus and nitrogen, but silver and gold"; this also true for watershed management. An estimated budget for the project implementation segmented into annual periods is needed. It might be helpful to organize resource allocations along the same lines as source assessment and linkages and document them. The other approach is to allocate resources according to function.

The plan must include all the costs associated with the proposed effort, such as the costs associated with the implementation of management practices, including all tasks and materials for the design, installation, operation, and maintenance of the management practices. The costs associated with the monitoring and evaluation as well as information and education components must be included. In addition to these project costs, an implementation support budget is needed. These are the administrative and management costs associated with every watershed management effort. The budget should differentiate between cash and in-kind contributions by the watershed management partnership.

7.3 PLAN FORMAT

The Ohio EPA developed a template for watershed action plans that can be used as guide for developing a watershed management plan document.[2] Table 7.1 presents the modified Ohio EPA template.

7.4 FUNDING

Partnerships need to look for diverse funding sources. Many funding sources are designed for specific types of activities. The partnership will have to piece together funding sources in order to meet the overall watershed management plan goal. Many cost decisions are based on what is affordable at the moment, not on long-term cost–benefit ratios.

Most partnerships start by looking locally for funding support and contributions. Many government grants require local matching funds or in-kind service. Private foundations are often more flexible, but may favor groups that can attract several funding sources. It is important for the partnership to be aware of the administrative requirements for any funding received and the time schedule for receiving the funds.

In some circumstances, the partnership may want to develop a dedicated funding source to support the watershed management implementation process. The dedicated funding source could be either a stormwater utility or a special-purpose district. Because of their watershed perspective, stormwater utility districts can develop programs and allocate resources across municipal boundaries to achieve water-quality goals throughout a watershed. A stormwater utility is essentially a special assessment district set up to generate and manage funding specifically for stormwater management. Users within the district pay a stormwater fee, and the revenue generated directly supports maintenance and upgrade of existing storm drain systems; development of drain systems, drainage plans, flood control measures, and water-quality programs; and supports administrative costs and sometimes construction of major capital improvements. Unlike a stormwater program that draws on general revenue funds or uses property tax for revenue, the people who benefit or contribute to the problems are the only ones who pay into the utility. An upside to a stormwater utility concept is that it provides a degree of fairness lacking in tax-based systems.

TABLE 7.1
Template for a Watershed Management Plan Document

1. Define the Watershed.
   Name, size administrative boundaries of watershed.
   Geographic location, USGS-HUC (8 digit minimum), River Reach or lat/long of the furthest downstream point.
   Background/historical information on previous efforts, etc.
   Purpose of plan — what will be achieved and why.
   Scope and limitations of plan — comprehensive, pollutant specific, etc.
   Who was involved in preparing the plan.
   Outline of the plan's content.
2. Describe the Watershed.
   Natural features/characteristics of watershed.
      Special values
      Hydrology
      Land-use
      Existing point sources (schedule for re-issuances)
   Water Quality status for all water bodies (lakes, streams, wetlands — for ground water aquifer specific not segment).
   Use designation/attainment of segment
   Causes of non-attainment by segment
   Threats to non-attainment by segment
   List Water Quality Goals — WQS needing to be achieved.
   List of Natural Resource Goals — Prime farmland, fisheries, etc.
3. Identify Problems, Indicators, and Measures.
   Identify sources of contaminants and quantify loads to meet water quality goals (point and nonpoint source load reductions must be identified) — what will be the indicator and frequency of measurement.
   Critical area of definition and map by pollutant source.
   Habitat conditions — what will be measured, why, and how.
   Species of interest — what will be measured, why, and how.
   Other natural resource issues/concerns (what will be measured, why, and how).
4. Implementation Strategy including Monitoring and Outreach Component. Look to see if they are focusing on the maximum pollutant strategies of a rotation, rather than planning on the average to get an average environment, looking at off-field planning — stream banks, in water body restoration and management, etc.
5. Document Planned Activities.
   Describe specific management objectives, approaches, and actions by category.*
   BMPs available with operation/maintenance requirements.
   Point Source Permit activities needed.
   Link actions with roles and responsibilities.
   Match activities with indicators and measures (defined in 3).
   Milestone schedule (Gnatt Chart).
6. Budget.

* Implementation, Information and Education, Monitoring/Evaluation, Administrative.

REFERENCES

1. Canfield, D., Welcoming address to the North American Lakes Management Society, Cincinnati, 1992.
2. Ohio Environmental Protection Agency (OEPA), *A Guide to Developing Local Watershed Action Plans in Ohio*, OEPA, Columbus, 1997.

8 Implementation

Even if you're on the right track, you'll get run over if you just sit there.

— Will Rogers

8.1 INTRODUCTION

The watershed management plan provides the implementation framework, which includes management practices, outreach, monitoring, and evaluation activities. It also includes all the activities necessary to execute the management plan and achieve the goals and objectives. Partners need to dedicate human and financial resources to support the various implementation activities depending on their interest, resource availability, and authority. Coordination and cooperation among partners are essential for successful implementation.

The partnership's operation committee forms teams to support the implementation of the watershed management plan. The implementation team approach has many benefits for the watershed management effort: it builds confidence in unbiased and equitable installation of BMP systems regardless of property ownership; it serves as a way to ensure that all agencies and organizations use the same standard of judgment; it allows different agencies and organizations to be the primary point of contact with various target audiences; and it expands the resources available to follow up with participants. Putting together a competent team to support the plan implementation is important for the partnership. To ensure the maximum efficiency of the implementation efforts, private consultants, vendors, and other service industry personnel should be brought into the implementation process when practical. Implementation teams can be structured around geographic areas, pollution sources, target audiences, and activities. Team makeup is based on the type of tasks or activities it is assembled to undertake; it takes many types of expertise to implement comprehensive watershed management efforts. Each implementation team needs to have a point of contact. The team's point of contact is responsible for scheduling and reporting.

Potential participants within the watershed need to receive clear messages about the proposed efforts, its purpose, and its value. Implementation teams have to know exactly what the partnership is trying to achieve. This means knowing the outcomes desired, the available tools, and the targets that must be achieved along the way to ensure the watershed management plan's goals are realized. Also, everything the team does needs to be focused on the implementation effort. Implementation teams can use a combination of approaches consisting of technical assistance, financial incentives, and outreach efforts to influence stakeholder behavior to achieve the necessary level of implementation to meet the water-quality goals.

Implementation teams should utilize a "master schedule" approach to increase the efficiency of their efforts. A master schedule lays out each task, dates to be completed, resulting products, resources needed, responsible persons, and what will be measured (Table 8.1). Having a master schedule increases the likelihood that the solution will be implemented effectively and on time. The master schedule approach also prevents partners from underestimating the time and resources needed to get a task done. In addition, the approach indicates when complex tasks need to be broken down into parts. Master schedules may be included in the watershed plan or developed just for use by the implementation teams to assist in their efforts at the lower levels. An effort should be made to prioritize activities, execute them as effectively and efficiently as possible, and document success. Implementation of an action should proceed at any point that is possible, legally supportable, and makes sense. Typically, if the watershed management effort is demonstrated as producing results and benefits, additional funding can be acquired. Master schedules are extremely useful in progress reporting.

The plan's objectives for land treatment, or implementation of pollution control technologies, must be related to the plan's water-quality goals and objectives. The plan should provide an idea of the type and extent of activities required to produce a change in water quality. This includes a description of measurable milestones for determining whether (1) management practices, other control measures, and restoration techniques are reducing the pollutant of concern, and (2) progress is being made in achieving the plan's goals and objectives.

Implementation efforts need to demonstrate results early. Implementation teams need to pick activities that will provide visible results and are appropriate for the schedule. Then do the activity and promote it so the public will see the results. Successful activities, even small ones, get the public and stakeholders interested and inspire them to do more. They also show potential financial supporters that the partnership can get things done. The implementation teams need to link this early demonstration effort to the communication plan to give project participants and cooperators good publicity. For example, the water-quality improvements installed as part of the Double Pipe Creek RCWP and subsequent publicity greatly improved the agricultural community's image in the eyes of the county government and public.[1] Publicity is especially important when a local business has allowed its employees to participate during work. Public recognition is a great thanks for businesses, organizations, and agencies. Look for all types of support from partners and possible contributors. Sometimes businesses and local governments find that providing in-kind services or raw materials is easier than funding. Partnerships need to make it easy for contributors to participate by looking at the partnership's request from contributors' perspectives.

8.2 INSTITUTIONAL ASPECTS OF IMPLEMENTATION

Besides the master schedule approach, implementation teams must work closely with the planning committee and evaluation efforts to track implementation progress.

TABLE 8.1
Master Schedule

Objective	Activity Being Implemented	Action Steps	Responsibility	Time Frame	Resource Estimate
Reduce sediment from eroding road-stream crossings	Retrofit existing culverts with extensions, reshape and vegetate side slopes	Change design criteria to accommodate current stream	Lead agency and road commission	Short-term	$2,500

DATE: _____   TEAM POINT OF CONTACT: _____

Developing an uncomplicated method for implementing management practices to decrease the amount of paperwork required while still allowing a minimum level of tracking is most important. The close working relationship between the implementation and evaluation teams is needed to ensure that necessary flexibility exists in the implementation efforts to modify the approach based on progress achieved and that new information is utilized as appropriate. In the South Dakota RCWP the project's implementation goals were modified after the results from the intense monitoring and implementation efforts indicated that animal operations were an important pollutant source.[2]

The implementation team must pass pertinent information from its experience to the planning committee, for it to use to modify the watershed and subwatershed plans. In this way, the local implementation teams are more likely to have workable, proven procedures and criteria included in the subwatershed plan as guides for future implementation efforts. Where subwatershed plans do not exist, assessing overall cumulative impacts of land use on water and related natural resources is difficult. In these situations, techniques should be used to minimize, to the extent possible or practical, the impacts on water and related natural resources at the parcel level.

8.3 IMPLEMENTATION SCALE

There are three scales of implementation: watershed, community, and parcel levels. The watershed-wide implementation is usually focused on stewardship and pollution prevention activities supported by outreach efforts. Figure 8.1 shows the scale and nesting of the various levels of implementation.

8.3.1 COMMUNITY SCALE

A community-wide focus requires analysis of drainage basins within their jurisdiction. Some watersheds may lie entirely within community boundaries. Other watersheds that are partially within the community boundaries must be included. While implementation is limited by jurisdiction, the technical analysis must take into account the hydrologic contributions from adjacent parcels within the watershed. Some problems are specific to a small waterbody where the entire watershed lies within the limits of a given community. In these small watersheds, an activity might involve citizens in just one community, who work with the implementation team to develop action items within the political jurisdiction. More commonly, however, the challenge with addressing watershed problems is that the watersheds do not conform to political boundaries and so multiple communities must be involved and coordinated to achieve watershed goals. Intermunicipal agreements, an available tool, work best on a case-by-case basis.

Implementation at the community scale usually falls into one of two categories: delegated or desired. To assist with implementation within a jurisdiction, a cover letter from the steering committee highlighting recommended actions for the specific municipality is an important introductory step. The cover letter should detail how implementing these actions could work with the ongoing municipal programs and priorities. The delegated or sometimes called "mandated" implementation is a result

Implementation

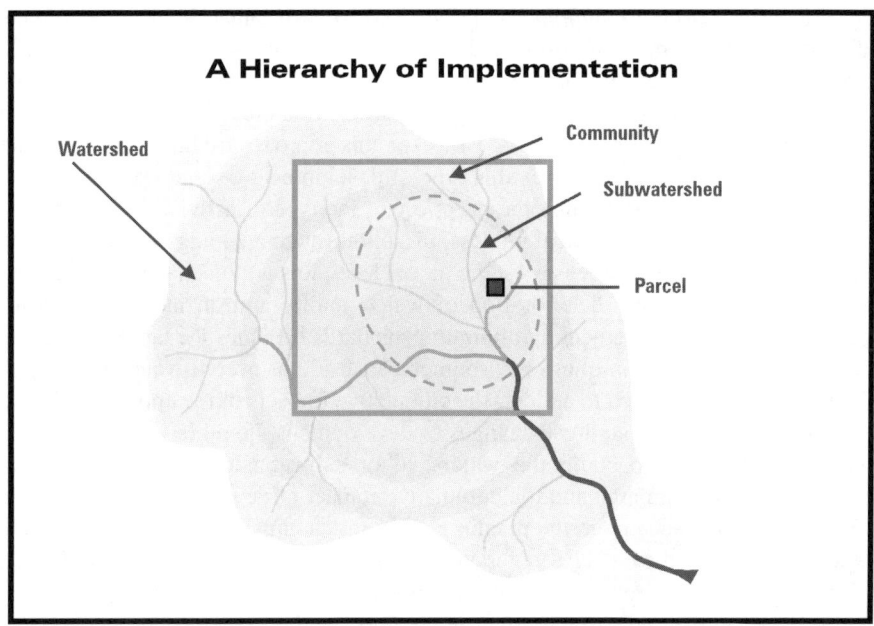

FIGURE 8.1 A hierarchy of implementation: watershed–community–subwatershed–parcel. (From EPA Office of Water, *Protecting Natural Wetlands, a Guide to Stormwater Best Management Practices*, pp. 4–9, 1996.)

of a community's being responsible for the implementation of a federal, state, or tribal program within its jurisdiction. Delegated programs are required by law, or the community assumes it is in their best interest. The community scale is the traditional level for implementation of nonstructural and management techniques to reduce the potential for future pollution associated with development and redevelopment activities. This requires communities to utilize municipal ordinances to provide the standardized regulatory framework to ensure the activities are implemented consistently within their jurisdiction. The watershed management planning process provides the opportunity for communities to enhance the implementation of their delegated programs. Implementing an overlay district through zoning is a way to provide some areas greater protection than others for water quality. An example of this approach is the Lake Whatcom management effort, a tributary to Puget Sound and a drinking-water source for the city of Bellingham, WA. Lake Whatcom needed protection due to development pressures. In 1999 Whatcom County adopted an overlay zone that limited land use and created stricter development standards. Designating Lake Whatcom as a special district enabled the county to impose additional requirements under the stormwater and land-clearing ordinances. Additionally, the county adopted an ordinance to transfer development rights to growth areas outside the lake watershed. In 2000, for Bellingham's portion of the watershed, new rules were enacted limiting the types and sizes of developments, prohibiting land clearing and grading on areas greater than 500 ft^2 between October and April, and restricting impervious surfaces to 2000 ft^2 or 1% of parcel areas (whichever was greater).[3]

Presently, most communities use regulations to ensure that the negative effects of land-use development are mitigated or minimized. This is usually done through a regulatory process enacted at the community level and implemented at the site or parcel level. This is viewed as enforcing restrictions of a prescribed nature on a site to minimize on- and off-site impacts. However, this approach usually avoids the first question that should be asked, "Is this type of development or land use appropriate at all?" This aspect, questioning the propriety of the type of activity on a particular parcel, is managed by a political process, usually through a zoning board. Since the decision-making process focuses on the parcel level, this approach has continued to result in an incremental deterioration of water quality and an increase in water quantity because management by minimum standards becomes the target for design that in turn ensures a minimum environment. Instead, an overall watershed framework should be employed to address the cumulative impacts of the approach. Where possible, any proposed land-use changes or development should be evaluated based on their potential impact on the watershed or subwatershed system, including upstream/downstream area and the cumulative impact of development activity. The watershed-zoning scheme in the planning chapter is a cumulative impact approach. It does not address the paradox that identification of a source does not equal location of the source. For example, the plan identifies the source-impervious area but does not locate it on the landscape to provide direction to the implementation team.

Desired programs are those the community adopted in response to a need it identified and wanted to address. Desired programs are typically associated with a watershed management plan. In these cases the program is narrow in scope or purpose and relies on following a specific process or management practice approach to solve the issue addressed. An example of this type of program is the Conservation Development Evaluation System (CeDES), a rating system created by The Conservation Fund to evaluate a conservation development over the development's lifetime with an emphasis on the water quality and landscape impacts. The purpose is to have developers think about environmental concerns earlier in the planning process and to give communities a means of assessing the impacts of better site-design practices. CeDES looks at conservation development in four specific areas: site design and construction practices; stormwater management; preservation of open space; and protection of natural resources. CeDES is intended primarily for areas of the Eastern and Midwestern U.S.[4] As a feature-oriented system where points are awarded or deducted for satisfying specified criterions, the scoring system is based on the premise that developments should always meet certain basic standards; developers can adjust their designs to meet a certain point total.

No single urban management practice is sufficient when used alone — and when used in combination with other management practices, their benefit and performance are interrelated. Taylor[5] suggests that a comprehensive and effective urban runoff management program must include both structural and nonstructural management practices in combinations designed to remove pollutants of concern. He suggests that urban planning and design efforts need to address peak flow and volume increases as part of the pollution control efforts. An assessment of the watershed is necessary to determine the appropriateness of each management practice for specific conditions (soils, topography, drainage area size) in the watershed. The assessment is also useful in establishing preliminary designs for the various management practices and in developing performance standards with which to evaluate the effectiveness of any implemented practice.

8.3.2 FIELD SCALE

The field or land parcel is the typical planning and implementation unit for the private rural landowner/operator or public landowner. Field-level analysis needs to take place within an overall watershed/subwatershed framework. Usually the field will not coincide with one runoff area. Occasionally, all the water on a parcel is generated from precipitation on that parcel. Often the water budget for a parcel is a combination of precipitation and runoff from adjoining parcels. Analysis at the parcel level is needed for the development of site-specific control strategies. The site-specific control strategies will have a design storm of either 10yr/24hr or 25yr/24hr. The parcel-level analysis needs to focus on hydrology and the pollutant delivery process; availability, detachment, transport, and discharge to a waterbody or conveyance system. The analysis needs to be able to identify opportunities to either disrupt the pollutant delivery process or eliminate it entirely.

Parcel-level analysis is extremely important within critical areas; this is to ensure identified sources are adequately addressed. For parcels extremely disturbed or where land-use conversion has recently occurred, the implementation team must perform a thorough historical analysis of the parcel. What is underneath could limit alternatives and will affect implementation costs from excavation efforts to soil amendments. The parcel-level analysis focuses on the control of four primary factors: water runoff and soil moisture, erosion, nutrient loading, and contaminant loading. These factors are dependent and highly interactive. Runoff control, for example, helps reduce sediments, nutrients, and pesticide loadings to waterbodies nearby. Where parcel-level analysis is not possible due to staffing or financial constraints, partnerships must establish a minimum level of management by land use. For example, on agricultural cropland a minimum level of acceptable implementation is a CORE 4 system.

CORE 4 is a common sense approach to improving farm profitability while addressing environmental concerns. The universal use of structural management practices is very expensive and, unless it provides a realistic level of protection for receiving waters, its use could be a total waste of investment. The approach is easily adaptable to site-specific characteristics and landowner constraints. Conservation buffers, weed and pest management, crop nutrient management, and residue management are the four core management practices in this system approach. This approach works best when it is integrated into the overall farm management plan. CORE 4 is economically beneficial to landowners, and so economic aspects of management practices are a major issue with participants. The Double Pipe Creek RCWP found farmer participation was affected by economic conditions, current government program requirements, and farmer-agency relationships. The project found implementation of management practices was correlated to farmer income, which is difficult to predict over a 10 to 15-yr planning period.[1]

For urban areas it is in the interest of site planners at the local level to consider site management on a systemic basis and reduce the impervious area. The systematic approach to addressing imperviousness is based on establishing maximum limits on impervious area in a subwatershed and then using parcel analysis to look for opportunities to limit imperviousness to, or below, the established limits. The watershed-planning process provides the opportunities and mechanisms for "trading" impervious area between parcels and subwatersheds.

Based on the parcel level of analysis, the implementation team sets management goals for fields, loading reductions, taking into account watershed management plan goal(s), on-site resource protection goal(s), field location, and landowners' objectives. The implementation team then designs a management system to achieve these goals. The selection of management measures for achieving the necessary loading reductions must factor in the site-specific characteristics of the parcel, existing management operations, and both the natural variability and the difficulty in predicting the performance of the management measures over time. Implementation teams need to allow sufficient time for designing and installing labor-intensive practices. For management at the parcel level, the overall goal is to minimize that parcel's maximum pollution-generating aspects. For example, there is an issue related to the development of site-specific management systems for agricultural cropland sources. Site-specific designs are based on the crop rotation for a particular field. Instead of the traditional approach of using the entire crop rotation in developing the management system to address average conditions for a particular field, a pollutant risk reduction-based management system needs to be designed. Agricultural pollutants carried in runoff from fields managed under multiple-stage crop rotations will vary by crop stage of the rotation, so a management system that minimizes the generation and transport of pollutants during the maximum pollutant production stage of the rotation needs to be designed and implemented. Table 8.2 compares the various management systems possible; for the corn portion a rotation indicates that switching to a no-till system will substantially reduce erosion and runoff. This creates a margin of safety in the decision-making process for natural variability and the difficulty in predicting the performance of the management measures.

Another aspect of implementation that needs attention is the scope of the parcel or site-specific planning of pollutant management systems in riparian areas. The Sycamore Creek (Michigan) Project documented that existing land treatment approaches must increase the focus on edge-of-field practices to improve water

TABLE 8.2
Effects of Tillage Systems on Soil and Runoff

Tillage	Soil Erosion (tons/ac)	Water Runoff (gal/ac)
Moldboard plow	7.3–23.1	43,700–87,600
Ridge till	1.4–10.1	21,400–58,000
No till	0.5–1.6	23,500–40,000

Source: Davenport, T.E. and Kirschner, L.T., Landscape approach to sediment control, 7th Federal Interagency Sedimentation Conf. Proc., Reno, 2001.

quality. The project documented a 60% reduction in suspended solids concentrations in the Willow Creek subwatershed. A direct correlation existed between the extent of no-till cropland management and sediment reduction over time. However, Marshall Drain, which had a greater percentage of cropland treated with no-till management, did not show a significant reduction in suspended solids measured. The only difference between the watersheds was how stream-bank instability was addressed. The results indicated controlling soil erosion from cropland alone was insufficient to reduce suspended solids. This on-site effort needs to be supplemented by stream-bank stabilization, such as the case with Willow Creek and not with Marshall Drain.[6]

Specific management systems are proposed for each parcel. Any two or more management practices used together to control a pollutant from the same source constitute a management practice system. Whole farm planning is recommended where resources and schedule allow for it. By planning for multiple issues, management systems that address more than one issue can be designed and recommended, thus eliminating the need to work with the parcel later to solve another problem that single-issue site planning did not initially address. The management system prescribed for each parcel must establish a performance level of management for that parcel. Performance levels can be achieved through a single management practice or a system of management practices, depending on the level of performance required and the site characteristics. Management practices are typically applied as systems of practices because one practice rarely solves all the water-quality problems at a site, and the same practice will not work for all the sources and causes of a pollutant. Implementation teams should encourage participants to implement systems of management practices to meet performance criteria. Systems of management practices are more effective than a single management practice in controlling nonpoint source pollutants from critical areas. The selection of management practices for a management practice system is based on the type of pollutants, pollutant sources, mode of pollutant transport, geographic location, site-specific characteristics, as well as landowner's financial situation. The implementation of management practice systems is also referred to as land treatment.

All three types (vegetative, managerial, and structural) of practices may be needed in a system approach to solve water-quality problems. Each type of practice has specific aspects that need to be considered with designing a system. Structural practices require construction activities to install and engineering assistance to design. Vegetative practices use plants to stabilize eroding areas or to treat runoff. Managerial practices involve changing the operating procedures at a site.

The implementation team must prepare for aggressive follow-up. Most management practice systems that include structural practices need continued maintenance in their early years and as farming equipment is upgraded or aspects of the operation change. For vegetative practices it may be necessary to try many different plantings to see which works best with the farm operation and site-specific characteristics. The team will have to work with the participant to plant and replant until the proper vegetative canopy is established. Successful watershed management efforts are a result of an incremental learning process and continued improvement in O&M of installed practices.

8.4 MANAGEMENT PRACTICE ASPECTS

To assist in the selection of management practices, a number of management practice manuals are available. The references provide the most comprehensive listing of materials available on management practices. One of the best manuals available is the *Guidance for Specifying Management Measures for Nonpoint Sources in Coastal Waters*.[7] The USEPA is in the process of updating the contents of the manual's individual sections. The guidance manual promotes the concept of management measures to address or prevent nonpoint source pollution. Management measures are economically achievable measures for control of addition of pollutants, which reflect the greatest degree of pollution reduction achievable through the application of best-available nonpoint source control practices, technologies, processes, site criteria, operating procedures, or alternatives. Generally groups of affordable management practices implemented together in a system approach achieve performance levels specified in management measures. Requiring the management measures applied everywhere alleviates the issue of "palatable" management practices approach. The palatable approach is voluntary, and landowners implement only the management practices they prefer rather than what is necessary to management's on- and off-site impacts.

Common evaluation criteria for selecting management practices are as follows:

- Pollution effectiveness. How well a practice reduces the following parameters:
  - Sediment
  - Nutrient
  - Phosphorus
  - Runoff
- Capital cost. Costs that would be incurred by the entity to implement the management practices or control technology.
- Operation and maintenance requirements. Those costs required to keep the management practice working properly, for point source controls cleaning and replacement.
- Longevity/reliability. How long will the management practice or control techniques last?
- Confidence. How consistently a management practice works in reducing a problem; in many cases, the scientific evidence is not yet available to assess the confidence associated with a given a management practice.
- Adaptability. Its use in various geographic areas and situations.
- Potential side effects. The possibility of causing another problem by treating the problem of immediate interest.
- Concurrent land management practices. What other management practice is it compatible with?

The current knowledge of management practices is expanding rapidly as more data become available. For example, data recently started becoming available for the effectiveness of rain gardens or bioretention cells in Largo, MD. A 456-ft^2 bioretention cell, which cost about $4500, was constructed to treat parking lot runoff.

The cell reduced metals (zinc and lead), phosphorous, and ammonia/organic nitrogen by at least 67% and dissolved copper by 50%.[8]

8.5 VOLUNTEER CONTRIBUTIONS

Volunteers can be very effective in assisting with implementation. Numerous activities of the implementation process are suitable for volunteer efforts. Neighbor talking to neighbor is one of the most successful participant recruitment techniques and follow-up for O&M activities for a partnership. Volunteers can also provide installation of labor-intensive management practices such as soil bioengineering and native plantings. It is important to note that the use of volunteers is not without cost. Training, equipment, transportation, and insurance can be required depending on the types of activities in which the volunteers are involved. Each of the activities carries a real dollar need that the watershed management effort or by a separate agency sponsoring the volunteer effort must meet.

8.6 IMPLEMENTATION FUNDING

Most planning meetings start and end with the question, "Where is the money going to come from?" The four-step watershed management planning process is "program-neutral," meaning the plan identifies the needed actions to manage the watershed regardless of the source of funding available. Once the plan takes shape, the planning committee, with assistance from agency technical advisors and local decision makers, starts looking for ways to fund its ideas. Getting the funds to implement the watershed management plan involves commitment, energy, and time, but the important elements for successful funding are already in place: organized partnership; systematic consideration of goals, needs, and alternatives. All are documented in the watershed management plan, which becomes an important element in the fundraising process. In addition to federal, state, and local government programs, partnerships can solicit funding from nonprofit organizations and private industries. For example, in the Oregon RCWP Project, the support of the local creamery was instrumental in obtaining high producer participation in the critical areas. The creamery had the ability to discount milk prices paid to producers who did not correct their identified pollution problems.[2]

8.7 IMPLEMENTATION DYNAMICS

The key to success for the implementation is to communicate. Even though the watershed management plan is a document reflecting consensus, conflicts among jurisdictions, agencies at various levels of government, partners, and the public are inevitable as implementation proceeds. These will need to be resolved, possibly by modifying the plan or the implementation mechanisms. The partnership chapter provides guidance on this aspect.

The implementation team needs to have a feedback loop to the operations committee and the overall partnership. This feedback loop is built on using land treatment tracking and water-quality data, honest sharing of data and information, and fine-tuning the management efforts based on progress and new information. The implementation team should schedule at least quarterly meetings with the operations committee to discuss progress, needs, and barriers.

For a complete listing of management practice guidance manuals, see Appendix 6-A, *Best Nonpoint Source Resources in Managing Lakes and Reservoirs*, North American Lake Management Society and Terrene Inst., 2001.

REFERENCES

1. Schaeffer, E.A., Farmer participation in the Double Pipe Creek, Maryland, in *Proc. Rural Clean Water Program Project, the Rural Clean Water Program Symposium, 10 Years of Controlling Agricultural Nonpoint Source Pollution: The RCWP Experience*, EPA/625/R-92/006, ORD/OW, USEPA, Washington, D.C., 1992, pp. 265–268.
2. USEPA, Evaluation of Experimental Rural Clean Water Program, EPA-841-R-93–005, USEPA, Washington, D.C., 1993.
3. Puget Sound Water Quality Action Team (PSWQAT), News from around Puget Sound, *Sound Waves*, 15(3), Fall 2000.
4. The Conservation Fund, The conservation development evaluation system, *Chicagoland Conservation Development News*, 1(1), 2000.
5. Taylor, S., Overview of conventional storm runoff water quality BMP characteristics and performance, *Stormwater Runoff Water Quality Science/Engineering Newsletter*, 2, May 19, 2000.
6. Michigan Department of Environmental Quality (MiDEQ), *Water Chemistry Trend Monitoring in Sycamore Creek and Haines Drain, Ingham County, Michigan, 1990–1997*, MI/DEQ/SWQD-99/085, MiDEQ, East Lansing, 1999.
7. USEPA, Guidance Specifying Management Measure for Sources of Nonpoint Source Pollution in Coastal Waters, EPA 840-B-92–002, USEPA, Washington, D.C., 1993.
8. PSWQAT, LID at work, *Sound Waves*, 16(4), Fall 2001.

Figure 8.1. USEPA, *Protecting Natural Wetlands, A Guide to Stormwater Best Management Practices*, EPA-843-B-96-001, USEPA, Washington, D.C., 1996.

9 Evaluation

There will come a time when you believe everything is finished. That will be the beginning.

— Louis L'Amour

9.1 INTRODUCTION

There is a dark side to evaluations. Think of your last performance evaluation. Evaluation is as critical to watershed management efforts as goals and objectives are. Most evaluations to document the project's achievement occur at the end of the project. Usually this is too late. Without collecting the appropriate information during the implementation phase, you cannot determine if the effort was successful in having its desired impact. Based on the proper design and data collection approach, evaluations need to be an ongoing part of any watershed management effort. The typical watershed management effort is designed to accomplish a specific task within a specific period and is usually evaluated against the completion of that task rather than the impact due to completion of the task. Kondolf and Micheli[1] state that despite increased commitment, evaluations have been generally neglected. An example of this neglect is the Section 319 funded North Fork Embarrass River Project (Illinois), whose purpose is to protect and improve the river's water quality by reducing nonpoint source pollution. The project will be evaluated against the number of installed management practices versus projected in the proposal, quarterly newsletters, bimonthly news releases, water-quality meetings and workshops, education/information tours and outdoor classrooms, and educational activities for schools. The project is not being evaluated against water quality or pollutant load reductions, so the project's impact in relation to the reason the project was funded is not even being examined as part of the project. In addition to accounting for the specific activities being started and resources expended, evaluations need to be designed to detect the long-term impact of these activities. Successful evaluation requires clear, meaningful, and measurable milestones and objectives for the plan and its implementation.

Evaluation is a process available for making watershed management partnerships more effective and efficient by providing a way to show others the value and role of various aspects of their efforts. Not every watershed management effort needs to be a research project to documenting cause-and-effect relationships, but some evaluation is essential for every watershed effort. When evaluating a watershed management effort, the partnership must systematically collect information about how the

141

efforts operate, and the effects it may be having on the actions of target audiences, changes in the environment, and what was expended.

Most evaluations of completed watershed management efforts are inadequate because they fail to detect if the project had the long-term impacts for which it was designed. Many efforts do not even complete an evaluation. In a study on river conservation enhancement, Holmes[2] found only five of more than 100 projects had postimplementation evaluations. The most common excuses for not carrying out evaluations are:

They take too much unproductive time.
They are of no value.
Circumstances were confounding.
Evaluations change as much as programs do.
There is no client for the results.
They are difficult.
The report often arrives long after the program is completed.
The process is too academic and complicated.

Other reasons cited are it is often unclear about the differences an evaluation makes, and it is presumed that the path to understanding achievement is both expensive and potentially disheartening. This path is called evaluation. Sometimes completed evaluations focus on descriptions of projects that are never carried out as designed or the original project design is substantially altered as the project is carried out. For example, under the President's Water Quality Initiative, the USDA established a five-year program to support a select few demonstration projects and the pilot targeting of technical, financial, and education assistance on a hydrologic unit area basis. The evaluation of this initiative was difficult because the effort was substantially altered as it unfolded, as the Cooperative Extension Services' budget was annually reduced from the original five-year commitment. It was estimated that most outreach efforts were supported at about 50 to 60% of the planned level, thus resulting in a substantial reduction in outreach activities (James Meek, personal communication).

Partnerships must be able to document what their watershed management efforts are achieving. Without such information, the existing funding could be reduced and volunteers may abandon the effort. This means that the steering committee needs to direct attention not only to establishing the goals for the watershed management effort, and how planning and implementation efforts are achieving the goals, but also at how and what will be measured. For example, rather than focusing on water quality, most watershed monitoring documents the number of management activities implemented and ends when the implementation funds have been expended. This approach fails to document the environmental impact of the implementation effort. In addition, the RCWP documented a lag time between water-quality changes and when management practices are implemented.[3] So linking project evaluation to management practice implementation creates an inherent bias toward failure to document water-quality improvements. This result is that most monitoring efforts

Evaluation

focus on the implementation process and the lag time between implementation and water-quality improvements.

Why evaluate? Who needs to know the partnership's results? Stating their expectations for evaluations is important for funders and decision makers so the partnership can meet them. For environmental issues the public and stakeholders need to understand what is causing the problem, what is being done, and if the management approach is working. Evaluation systems play an important role in linking management activities with the public and stakeholders and gaining their support. Through the evaluation process the problem is documented, management activities are accounted for, and changes to the environment are documented. Key audiences for evaluations are funders and decision makers. Evaluations provide a means of accountability. An essential aspect of accountability is the commitment organizations demonstrate to take the appropriate actions to seek out and openly report errors. The two types of accountability evaluations can address are (1) ethical accountability — partnerships must do the right things for the right reasons; and (2) economic accountability — partnerships are expected to do the best job achievable for the least cost possible.

Evaluation is critical to the long-term success of watershed management efforts. Evaluations allow partnerships to build on success, learn from mistakes, and modify implementation approach. Keep the following thought in mind, "Would you invest your own money and time in this effort, knowing that you would not get any feedback on performance of the investment?" Partnerships need to guide the evaluation effort so they can provide feedback to participants. Without an evaluation system for the overall effort, the partnership could be wasting time and money and not know it. The Highland Silver Lake RCWP (Illinois) Project is an excellent example of why evaluations are essential to a project's success. The evaluation effort showed the project planners had incorrectly identified the problem (excessive soil erosion) to be addressed. Implementation efforts focused on reducing excessive soil erosion and resulting lake sedimentation. This management practice implementation approach was ineffective in addressing the true problem, erosion of natric soils.[4] The best watershed management plan will need adjustments and tweaking throughout the implementation process. When the implementation teams find something is not working, they need to examine carefully why it is not working and then work with the planning and operations committees to modify existing implementation approaches. Once partnerships know what and why, then the plan can be adjusted as necessary. Partnership efforts are about solving the problems. If the steering committee has identified problems with the original management strategies and approaches, it will need to work with stakeholders to make sure it is adjusted accordingly. If the identified problems are with personnel, the steering and planning committees need to consider shifting or sharing responsibilities among partners. The important message is that it takes an evaluation to make the necessary changes to achieve the desired goals.

The evaluation process must account for social capital measures for communication, educational efforts, trust building, and conflict resolution. It also is important that the evaluation includes measures for institutional changes, economic impacts, and intermediate environmental changes. Most social capital measures require evaluators to conduct interviews, surveys, and document reviews. The breadth and systems orientation of partnerships can require complex evaluation designs that involve a mix of quantitative and qualitative methods and techniques to document cause-and-effect relationships.

Evaluations can show: changes in knowledge, attitudes, or awareness of issues, changes in behavior, which management practices were adapted and which were not, and changes in watershed conditions including water quality.

9.2 EVALUATION TYPES

Evaluations can be done any time and at multiple times. Specific evaluation methods, techniques, and approaches have been designed for specific purposes and each phase (planning, implementation, and postimplementation) of the watershed management process. While the framework for conducting evaluations is based on well-documented factors, many good decisions can be made through an evaluation guided by common sense, reality, and practicability.

The type of evaluation chosen will affect the following:

What information/evidence will be gathered?
Where will the information come from?
What tools will be used to gather the information?
How will we deal with matters of privacy?
How will the information be evaluated for results?
Who will analyze or review the results?
How will evaluation results be used?

A comprehensive evaluation will utilize a combination of evaluation types and methods to evaluate a watershed management effort accurately. For the purposes of this book there are four major types of evaluations: formative, process, outcome, and impact.

Formative Evaluation (Prior)

A formative evaluation is undertaken to test approaches, materials, and ideas. Additionally, formative evaluations are utilized to understand the target audience before a project is implemented. A key factor here is determining the appropriate target audience based on project goals and objectives. The needs assessment discussed in the social capacity building chapter is a type of formative evaluation. The principal difference between the traditional formative evaluation and the needs assessment is the decision that arises from the outcomes (see the social capacity building chapter for more information on needs assessment). According to Herman and others,[5] needs assessment results in the allocation of resources to meet priority needs. The formative evaluation is an attempt to improve implementation before getting started. For example, stakeholders and landowners are surveyed concerning barriers to adopting management approaches. Survey results are then utilized to develop an implementation approach that will overcome the barriers identified.

Evaluation

Process Evaluation (During)
While this type of evaluation is sometimes considered part of a formative evaluation for watershed management, it is a stand alone type that focuses on the tracking of activities and expenditures. "Bean counting" is another name for this type of activity. Monitoring and evaluation efforts are designed to assure performance of the management effort, to determine if the partnership is meeting its goals and objectives as measured through a series of milestones. Midterm failure to achieve milestones does not imply failure of the management effort, but instead offers an opportunity to understand better and then modify the management effort. Consider how many indicators are needed; most watershed efforts require the use of multiple indicators to account for complex processes and the uncertainty regarding individual indicator effectiveness. Environmental indicators are measurements of water quality, habitat, or other criteria that tell something about the environnment's health. Administrative indicators are beans that you can count. They are usually easy numbers to generate, but they are often intended as indirect indicators of the desired condition. The partnership should utilize an evaluation process that incorporates both environmental and administrative indicators in the ongoing tracking of the watershed management plan's implementation. Using process evaluation, the operations committee can monitor implementation activities and provide timely information to the steering committee, partnership, and sponsors. Process evaluation allows the steering committee to modify project activities in response to ongoing feedback. This type of evaluation helps partnerships correct errors, eliminate redundancy, refine the monitoring program, and test progress toward pollution control objectives.

Periodic reappraisal of the implementation strategies is necessary to see if changes would contribute to greater cost-effectiveness. Can cost be reduced or greater reductions in pollutant loadings be achieved? Are higher incentives needed for the more cost-effective management practices? A major problem in utilizing process evaluation in watershed management efforts is the inability of process evaluations to determine the cause of the problem (lack of implementation) and the lag time associated with reporting accomplishments. One crucial rule to keep in mind is that when the activity is not working, stop doing it — resources are too precious to waste. Data are needed to determine what is working and then trigger intervention for those activities that are not.

Process evaluations can do the following:

1. Monitor the planned actions by recording activities and number of participants systemically.
2. Keep everyone focused on the big picture.
3. Provide the database to allow the steering committee to evaluate cost-effectiveness (dollars spent/accomplishments) at every stage of the project.
4. Provide information for the communications plan.

The partnership needs to develop a series of milestones to measure a project's progress. Management practice, program, and activity milestones are three types of necessary interim or surrogate measures to evaluate progress. Watershed partnerships need to be familiar with the idea that implementation requires trial and error over time to learn how the water resource will respond to implementation efforts and what stakeholders are willing to do. Partnerships should modify their management approaches when milestones are not achieved. The Lower Big Rib Priority Watershed (Wisconsin) Project established an interim milestone to trigger such a modification. For this project if, after 5 years of implementation, the calculated sediment reduction, based on cost share agreements, is less than 60% of the total cropland reduction goal. Additional agricultural fields will be classified as critical.[6] Failure to achieve the interim milestone results in expanding the management effort into more cropland areas of the watershed. This expansion would increase the probability of achieving the necessary sediment reduction to achieve the project's water-quality goals. The ultimate goal of the evaluation process is to improve the ongoing implementation effort so it achieves the overall goal. Process evaluations focus on:

- Actions various partners will take to achieve milestones and goals.
- Program goals related to how things are accomplished.
- Status of objectives.

Examples of process evaluation coverage are:

- Tracking of management practice implementation (numbers, types, and location) and requests for technical assistance (numbers and types).
- Number of permits issued with new limits.
- Number of point sources in substantial noncompliance.
- Elapsed time from identification of permit violations to correction.
- The master schedule provides the documentation on how the evaluation will be conducted.

Outcome Evaluation (Afterward)

This is also called a summative evaluation and measures the short-term results associated with a project. The big question, "What happened for the money spent?" is answered here. In this phase the partnership has to rely on "reasonable probability" instead of "cause and effect" when evaluating these efforts. The partnership needs to feel comfortable that the results can be verified with a reasonable probability that the effort will have the expected long-term impacts. These evaluations usually rely on readily available information and rarely measure real progress in water-quality or natural resource indicators. Outcome evaluations can be used to:

1. Measure changes in knowledge, attitudes, awareness, skills, aspirations, or behavior.
2. Determine if the project worked within the desired time frame. Were the milestones met?
3. Determine if the project goes beyond the desired effects.

While changes in the environmental condition locally will ultimately impact the larger watershed scale, the partnership must be able to document that results are going in the right direction. Approaches need to be focused on a targeted area and population and monitored if the partnership hopes to verify improvements at both the local and watershed scales. For example, to evaluate the impact of a livestock operation on a downstream lake, the monitoring program needs to include water-quality variables characteristic of livestock operations and applicable to lake ecosystems. Phosphorus and nitrogen are found in animal manure and can impact lake water quality and therefore should be included in the monitoring program. However, simply monitoring for nitrogen and phosphorus downstream of any livestock operation is not sufficient to determine cause-and-effect relationships, as both phosphorus and nitrogen can be attributed to other sources. Other variables such as land use and precipitation must be evaluated to determine which factors are linked to in-lake water quality.

Watershed management is not about the partnership, but it is about what stakeholders and the public do as a result of the partnership's overall efforts. Sometimes an assessment is needed of how the partnership is working and what the external effectiveness of their efforts is. Where possible, partnerships should survey project participants and people who were eligible to participate but chose not to become involved in the effort. This information can provide feedback on the effectiveness of information and education efforts and may be of value to management cycles. While individual activities are important, it is the overall results that really count.

Any complete set of benchmarks will include some that can be used to determine trends in water quality. To know whether the partnership has made environmental progress, the partnership must know where it started. That is why a basic assessment of the chemical, physical, and biological condition is fundamental to a partnership's evaluation program. The basic goal of the evaluation is to determine whether management approaches worked; did implementation occur as planned? Determine which management practices, pollution control measures, or other environmental improvement practices were installed in particular locations to solve certain problems. What was the effectiveness of the control actions implemented? Management practice goals are the next best thing to water-quality goals because it can be assumed that the implementation of appropriate management practices will result in improved water quality. Management practice goals need to be set for structural, vegetative, and management practices. Are water-quality standards attained or projected to be attained?

The sediment reduction pilot project in the Maumee River provides an excellent example of an outcome evaluation. The Maumee River watershed encompasses more than 4.2 million acres in Ohio, Indiana, and Michigan. Almost 80% of the drainage basin is in cultivated cropland. The water-quality problem is high sediment loading to Lake Erie and sedimentation in the Toledo Harbor navigation section of the river system. It is estimated that more than 90% of the annual 1,268,000 tons of sediment transported by Waterville (Ohio) is a result of crop production. An average 850,000 yd^3 of material must be dredged annually from the Toledo Harbor navigation complex. Based on the assumption that a reduction in sediment entering from the watershed could impact the amount of annual dredging required, a two-year pilot sediment source reduction project was initiated in 22 counties in the watershed. To be eligible the counties had to develop sediment reduction strategies. These strategies usually had three major components in common: an incentive program for landowners, demonstration projects, and information and education activities. Due to the short timeframe (two years), natural variability of the system, size of the watershed, and lag time between land treatment and resulting water quality, measurable reductions were not expected in short-term dredge quantities.

Based on reasonable probability three measures were developed to evaluate the effectiveness of the pilot project: conservation tillage trends, long-term sediment concentrations-based monitoring, and estimates of gross erosion. Conservation tillage in the Maumee River basin increased from 52 to 54% during the pilot study, while statewide there was a 3% decrease during the same timeframe. The project evaluators concluded sediment concentrations decreased by approximately 20% during the period. The evaluation correlated this reduction with improved agricultural stewardship. Overall, the gross erosion measure indicated the pilot project would not be successful in the long term. While tillage was increasing, sediment concentrations were decreasing; changes in a USDA commodity program and crop prices resulted in a net increase in acres planted in corn and soybeans. This change in cropland will result in an increase of gross erosion and a subsequent increase in the sedimentation of the navigation complex. The gross erosion analysis also indicated that the size of the watershed makes it difficult to determine the actual sources of majority of sediment, and the conservation tillage approach needs to be supplemented with conservation buffers and land retirement programs such as the Conservation Reserve Program (CRP) to reach gross erosion reduction goals. CRP is a production reduction program managed by the USDA to reduce the amount of land tilled and thus reduce soil erosion and sediment generation.[7]

Impact Evaluation (Much Later)

Impact evaluation is the most difficult type of evaluation to complete. It measures the long-term impacts of a management effort. This type of evaluation is extremely important for pilot and demonstration activities that the partnership or individual partners want to promote in the future. William Jordan III[8] presents a four-dimensional look at evaluating restoration efforts:

1. The product. Is the restored system ecologically accurate with respect to functions and dynamics as well as composition and structure?
2. The process. Has everyone with an interest in the project had a chance to participate? Were the right questions and ideas about ecosystem restoration asked? Were the appropriate restoration techniques used?
3. For the participants. Was the project an occasion for learning and for emotional and spiritual bonding?
4. The performance. What information, ideas, and values did the partnership convey to people who were directly involved in the project but who could benefit from it as an audience?

Evaluation criteria utilized must be able to distinguish between failures of watershed management science and failures of poor application of the science. This type of evaluation requires durability of project goals, objectives, and reporting. It is important to note that over a longer time, the perceptions and expectations of stakeholders may change. Finally, the world did not stop as the partnership's efforts progressed, so evaluators must determine if the project's goals and objectives have remained constant and relevant. The project is likely to change as it develops and is implemented. Long-term results usually vary from the short-term results, so impact evaluation is needed to measure the ultimate value of the effort. Review the achieved and expected results to see what insights on cost-effectiveness and benefit generation can be gleaned of value for efforts by the partnership. With this type of evaluation the partnership can answer:

Did anything change?
Did the water get better?
Did the partnership achieve its goals?
Can people swim again?
Was the money spent worth it?
Were the right controls put in place?
Is the waterbody still on the water-quality impairment (303(d)) list?

For example, a Clean Lakes Project Phase 3 Post-Restoration Monitoring study was conducted five years after the successful completion of a Phase 2 lake-watershed management program designed to eliminate watershed sources of pollutants to Lake Le-Aqua-Na (Illinois) and thereby improve in-lake water quality. Phase 2 of the project was judged a success after a summative evaluation indicated the project met its immediate goals and objectives. The phase 3 study, five years later, documented that the management practices were still in place; that one practice had been improperly installed and maintained; while the effectiveness of some of the management practices had decreased, they were still achieving their goals; and that the overall management program was still having the desired effect on the lake.[9]

9.3 EVALUATION BARRIERS

Lack of knowledge about watershed processes is one of the key barriers to successful implementation of watershed approaches. Good evaluations can answer important questions about the process and impact associated with watershed management efforts. A major reason for this lack of knowledge is that the evaluation process is not conducted properly. There are two distinct groups of problem evaluations: not comprehensive and inappropriate. The inappropriate evaluation is by far the worse; it happens when the focus is just on bean counting rather than the environmental impact of the management. By focusing on what happens, partnerships lose sight of the environmental impact of their effort and then they cannot tell stakeholders and funders what the expenditure of funds has meant to the community. When evaluations are not comprehensive, usually it documents what the condition of the natural resource is at the end of the effort but cannot tell why it is in that condition, or it if it is improving or degrading. The lack of an adequate baseline, an incomplete monitoring scheme, lack of money, inadequate data collection and analysis and confounding circumstances result in inadequate evaluations. In addition, partnerships' overzealous expectations sometimes lead to the misuse and inappropriate application of the information generation by the evaluation effort. Most evaluations end when land treatment ends, and experience from previous efforts has shown us that this is the wrong approach.

In addition to the barriers mentioned earlier any number of factors may limit the scope, direction, and success of an evaluation. Leeds and others[10] identify a number of factors that may influence an evaluation. Several of these are applicable to watershed partnership evaluation. The appropriate factors are listed below:

- Organizational structure/politics. Is the partnership set up to implement the watershed plan?
- Program leadership. Has the partnership bought into the evaluation, or is it just a funding requirement?
- Economics. Can stakeholders afford the solution?
- Project evolution. Have the watershed management plan's goals and emphasis changed since it started?

Each of these factors has implications for an evaluation. Partnerships need to examine the purpose of their evaluation in the context of these factors. For example, is the evaluation going to be used to reorganize the committee structure or membership? Is a consensus for this needed? Knowing which of these factors might have an impact on your evaluation will allow the evaluation team to ensure data collection approaches and methods account for them.

Other problems associated with the evaluation process related to the methodology are:

- Project decision makers are often handed answers to questions that someone else felt it was important to ask.
- The evaluation report often arrives long after the program is completed.
- Evaluations are directed to the wrong people.
- Evaluations change as much as the implementation approach.

9.4 EVALUATION LEVELS

A holistic perspective is needed when monitoring the performance of the watershed management effort. Still, monitoring should be focused narrowly on the fewest possible measurements or indicators that most efficiently demonstrate the status of the watershed and the success of the management effort.

Many different evaluation tools are available for watershed management efforts' evaluation. Since the focus of most watershed management efforts is on changing people's behavior, Bennett's hierarchy of evidence provides a framework for organizing the evaluation. In addition, the planning committee needs to view Bennett's hierarchy as another management tool. The first three components (inputs through target audience) of the hierarchy of evidence can be considered project inputs, and the last four components (reactions through results) can be classified as project outputs. This classification of components enables the planning committee, during midcourse and annual reviews, to evaluate the projected impact of varying the project

TABLE 9.1
Bennett's Hierarchy of Evidence

Level	Component
7	End results
6	Behavior change
5	Knowledge, attitude, skills and ability (KASA) changes
4	Reactions
3	Target audience
2	Activities
1	Inputs

Source: Beech, R. and Drake, A., *Designing an Effective Communication Program: A Blueprint for Success, National Network for Environmental Management Studies Program*, USEPA, Chicago, IL, 1992, p. 60.

inputs. Table 9.1 offers a brief summary[11] of each of Bennett's hierarchy's components, in ascending order.

1. Inputs. Project resources that are used to carry out the work. These resources include at a minimum funds, paid staff, volunteers, office space, and supplies.
2. Activities. Event occurrences, actions that are done to implement the effort, such as planning.
3. Target audiences. The stakeholder groups targeted for the various aspects of the watershed management effort.
4. Reactions. The target audience's reactions or views toward the proposed and implemented activities.
5. KASA changes. The awareness, knowledge, skills, and ability of the target audiences members that are needed to induce a behavior change and have an impact on the environment.
6. Changes in behavior. Participants' behavior changes through the social capacity building process.
7. End results. Results related to the project's goals and objectives developed when the original activities were planned.

While achieving stakeholder behavior change is the first requirement, the evaluation team then needs to know if it is sustained and if it had the desired impact. All watershed management efforts have objectives at several levels of the hierarchy. Activities and efforts are often expended to accomplish an objective at the level for which it was designed. Good evaluations are often designed to measure at more than one level of the hierarchy.

Evidence of a watershed management effort impact becomes stronger as the evaluation ascends the hierarchy. The lowest two levels (inputs and activities) provide little or no measure of participant benefit or environmental improvement. If the main focus of the evaluation is to increase the efforts' administrative performance, it is important to apply more evaluation techniques at the lower levels (three and under). Overall, watershed management effectiveness requires evaluations to be conducted at the upper levels of the hierarchy. Evaluations covering levels 4 (reactions) and 5 (KASA) provide an indication of whether or not the implementation approaches are working. KASA changes give an indication of potential management approaches' adoption. Partnerships can use process evaluation techniques as program management tools to keep the effort on track. The planning committee needs to verify what is important, keep track of progress, and use data on which to base midcourse corrections. Implementation teams need to know what they have to learn about stakeholders and their behavior to ensure a successful effort. However, the planning committee should not count on changes in knowledge and awareness as indicators that the implementation approach is reaching its targeted behavioral changes. Documentation of behavioral change is needed.

External accountability increases as the level of evaluation increases, whereas internal accountability follows the reverse trend. A good evaluation plan should prompt the steering committee to think about the different questions to consider at the beginning, midcourse, and after the formal aspects of the project. Evaluation results are of interest to a wide range of stakeholders, including financial supporters (taxpayers), health and education professionals, environmental interests, media and communications specialists, staff, volunteers, and the targeted and nontargeted public.

The level to which an evaluation is carried out has a tremendous impact on cost, requirements, and usability. The higher up the hierarchy you go, the greater the probability that external factors will influence the project results. Therefore, it is more difficult to design an evaluation that can detect these problems. The evaluations that utilize the higher levels of the hierarchy usually are more expensive because of data collection requirements and increased time to obtain results. Additional expertise is required to design evaluations, analyze data, separate external influences from an actual project, and provide feedback. For demonstration and pilot efforts the cost for higher-level evaluations can pay for themselves in terms of increased efficiency and effectiveness of programs and efforts developed based on the results of the pilot.

Evaluation techniques selected for higher levels of the hierarchy must be flexible enough to respond to change. The evaluation process must always link the plan's goals with the impacts and any needed modification. For example:

- Plan goal. Reduce sediment delivery loading to Emily Lake by 40%.
- Management question. Can sediment delivery to Emily Lake from a new development be reduced to an acceptable level?
- Monitoring objective. To evaluate the effectiveness of proposed management practices in reducing sediment delivery from the new development.
- Monitoring results. Management practices effectively reduced sediment delivery to the identified level, with buffer strips providing the greatest benefit; however, efforts to establish a three-zoned buffer were not successful due to poor survival of shrubs and trees.
- Management improvement. Use a similar management practice package at future sites, with an emphasis on three-zoned buffers. Modify shrub

TABLE 9.2
Modified Bennett's Hierarchy for Watershed Management Effort

Level	Component
7	Behavior and resulting environmental change
6	End results (linked to funding period)
5	KASA changes
4	Reactions
3	Target audience
2	Activities
1	Inputs

and tree recommendations to favor hardier species. Adjust O&M requirements to increase survival rates of plant species.

The way Bennett's hierarchy has been traditionally applied makes it inadequate to document the true impacts of watershed management on long-term water quality. However, experience shows that a modification of Bennett's hierarchy of evidence for program evaluation is very applicable to watershed management efforts. Table 9.2 presents the hierarchy modified for watershed management. This modification shifts the focus of the evaluation to long-term behavioral changes after the planned management activities have been completed.

Reversing the order of changes in behavior and end results is more appropriate for watershed management efforts. The end results evaluation would focus on surrogates for water quality and completing the implementation plan. This modification will allow evaluators to determine if the necessary O&M is occurring after the initial implementation period and if water quality has improved. It also takes into account the "Hawthorne effect"[12] when participants respond favorably (implementation) to attention rather than being committed to change, and once attention has faded they revert to their previous behavior. The modified hierarchy attempts to link long-term behavior change with long-term changes in the environment that are associated with the behavior in question.

Table 9.3 shows that the number of hierarchy levels varies by type of evaluation. Due to cost and complexity, the steering committee must decide what type of evaluation is needed to document its overall project or each phase. While not all watershed management efforts need the same level of evaluation, some evaluation is needed. Any evaluation should be planned prior to initiation of implementation activities — not as an afterthought. The level and type of evaluation utilized must be linked to the partnership's needs and the effort's goals. For example, if implementation status is needed, a process evaluation that focuses on levels 1 and 2 of the hierarchy would be sufficient. However, if the steering committee is contemplating changing the direction of the plan or enhancing specific components, the evaluation must be able to correlate data from levels 1 and 2 to the target audience (level 3), reactions (level 4), KASA (level 5), and end results (level 6) for the existing

TABLE 9.3
Modified Bennett's Hierarchy as It Corresponds to Various Evaluation Types

Level	Component	Formative	Process	Outcome	Impact
7	Behavior changes		X	X	X
6	End results		X	X	X
5	KASA changes		X	X	X
4	Reactions		X	X	X
3	Target audience	X	X		
2	Activities	X	X		
1	Inputs	X	X		

efforts and what would be expected with the future efforts. This type of effort would require a more comprehensive evaluation than just tracking status. For example, the Lower Big Rib River Priority Watershed Project's evaluation plan includes an administrative review, a pollution reduction evaluation, and water resource monitoring. The administrative review of the project includes information on accomplishments, financial expenditures, and staff time spent on project activities. Nonpoint source pollutant reduction goals will be compared against reductions estimated on land treatment activities. The water resource monitoring consisted of two types of evaluation: monitoring conducted statewide, and on "signs of success." The "signs of success" monitoring is a short-term effort designed to provide early evidence that better land management does make a difference.[6]

Table 9.4 shows what level of the modified hierarchy should be addressed by the watershed management process phase. Assessing watershed residents' needs and behavioral changes at several points in time provides an indicator of program effectiveness. Partnerships can use Table 9.4 to design evaluations to build from one phase and scale of the process to the next. For example, the parcel scale is a cost-effective way to document off-site water-quality changes associated with a management approach. For the implementation phase, evaluations help guide the implementation approach. An ongoing evaluation process determines:

TABLE 9.4
Modified Bennett's Hierarchy as It Relates to Process Phase

Level	Component	Planning	Implementation	Evaluation
7	Behavior changes		X	X
6	End results		X	X
5	KASA changes		X	X
4	Reactions	X	X	
3	Target audience	X	X	
2	Activities	X	X	
1	Inputs	X	X	

Evaluation

If the existing implementation approach is solving the wrong problem

If the partnership is solving one problem only to find another, more difficult, problem being masked by that the pollutant of concern

If the implementation approach is not effectively reaching the goal

If the project is meeting some administrative goals but not the water-quality goals.

If the partnership set the wrong goal, or the goal was too low to solve the problem.

These potential problems are standard complications for any watershed management project with extensive land-use management activities. Based on the results, the partnership can go back to the planning phase to design and conduct studies to redefine a water-quality problem or adapt management approaches to be more effective. The four-step process is flexible enough to support the adaptive management approach. Adaptive management or iterative implementation needs to be based on information (Figure 9.1). Adaptive management requires the partnership to use management tracking data and available monitoring information to make modifications to the implementation approaches being utilized. Decisions regarding modifications to the implementation approach must be based on the best information available, and sometimes the decisions will not be completely right. This reality requires the project to build this opportunity into the annual progress evaluations. If the evaluation identifies problems, the implementation team needs to work with

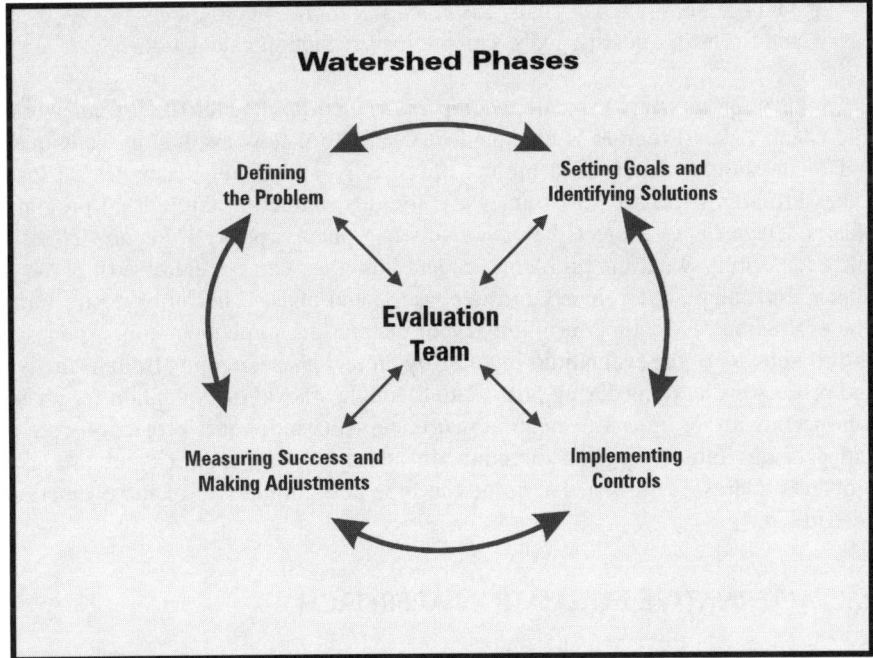

FIGURE 9.1 The four watershed phases.

the steering and planning committees to change timeframes, objectives, or implementation approaches to put the project back on schedule. Regular evaluations can help identify problems and concerns early in the watershed management process. Terrene[13] recommended that the annual partnership evaluation have different groups of stakeholders review each portion of the project using the same evaluation criteria to avoid bias. The evaluators attempt to reach a consensus on overall management approach, list of activities that must be modified, and a master schedule to implement it. The implementation and monitoring teams need to coordinate their activities to ensure sufficient information and data are available for the evaluators.

This allows partnerships to examine the tools and materials they have developed and used to support implementation. Partnerships should include in their evaluation plan activities that cover product utilization as well as implementation progress. At a minimum, watershed management evaluations should consist of measuring the achievement of administrative goals (management practice, activity, program) and monitoring the observed and documented environmental results of the project (people's behavior and associated management practices). Documenting administrative accomplishments is important since showing actual environmental improvement can take years. Partnerships need continual feedback. Measures other than water quality and natural resource condition can be used to document interim progress. Terrene[13] provides a list of methods that can be used to measure particular goals:

- Program goals. Periodic written reports, public meetings, financial records.
- Activity goals. Tracking and reporting systems to report activities.
- Management practice goals. Lists, maps, reports, photographs.
- Interim water-quality goals. On-site load reduction calculations..

Some programs have evaluation components incorporated into their framework. The Clean Lakes Program is an excellent example of how evaluation techniques and focus should change as a project progresses. The Clean Water Act's Clean Lakes Program (Section 314) framework stresses evaluation through all program phases. Clean Lakes Projects have three distinct phases: phase 1 diagnostic/feasibility (determine what the problems are and how they can be addressed); phase 2 implements the plan developed during phase 1; and phase 3 performs postrestoration evaluation on the long-term impacts of the phase 2 implementation. The level and complexity of the evaluations increase by project phase, ranging from formative and process evaluations during phase 1 to the highest level of evaluation for phase 3 projects with an impact evaluation that documents individual project outcomes and provides future program direction. In addition, the complexity of the lake problems addressed further determine the type and complexity of the evaluation needed.

9.5 ALTERNATIVE EVALUATION APPROACH

Very few partnerships evaluate the impact of their work; rather, they document activities accomplished. Many partnerships believe they have contributed to improvements but cannot document any links with the actions. Another approach to documenting the impact of a watershed management effort is to examine the value added by the effort. An example of this approach is RESULTREACH.

RESULTREACH is a way of reaching over implementers and representatives to the stakeholders customers presumed benefit by the programs. RESULTREACH seeks not to establish causality or absolute proof of anything. It is, rather, a reasonable way of establishing the "value added" of a given program or intervention, and it is a tool for improving the partnership's efforts that clearly suggest just how to make them more value-adding to stakeholders. RESULTREACH shifts the focus from what implementers do to what stakeholders gain. In most cases, RESULTREACH finds that the way stakeholders define, use, and gain from programs is quite different from how implementers presume to do so. RESULTREACH is a highly focused tool that does not intend to do everything included in a comprehensive program evaluation. RESULTREACH is far more straightforward and less expensive than a full and formal evaluation. Further, it can more readily be placed in a context of learning (e.g., to improve investments in programs and program operations) than of evaluation (as judgment or assessment). While RESULTREACH does not represent itself as scientifically valid (e.g., it makes no pretense of proving causality), it does provide a reasonable level of confidence in its findings. It goes well beyond anecdotal evidence and avoids the many traps inherent in many forms of numerical data, including surveys.

RESULTREACH implementation for a watershed partnership has these general steps.

1. Set expectations. Develop with the partnership involved a clear set of questions to which it wishes to gain answers from RESULTREACH, and the level of verification standard it desires to see met.
2. Select stakeholders. Defining a set of program beneficiaries to be sampled typically starts with a full list of program participants and a random sample drawn to reflect proportions of people by at least two stakeholder factors known to be associated with differences in outcomes.
3. Clarify stakeholder access. Define a way to communicate (e.g., phone numbers or addresses) with the stakeholders selected. In addition, resolve questions of privacy and confidentiality.
4. Conduct the reach. Stakeholders are each asked a series of questions, often by phone, occasionally in person, but always one at a time.
5. Analyze the responses. Stakeholder result data are now carefully assessed for patterns, trends, and many forms of coherence as well as of divergence.
6. Ensure learning. Now the evaluator works with the partnership on developing a strategic response to the information and patterns found in such areas as product improvement, calculation of return on investment, and consideration of alternative investment and implementation approaches. Investment, in this case, is the technical and financial assistance; the outreach efforts are for product evaluation.

9.6 EVALUATION PLAN

Several guiding principles exist to conduct watershed management program evaluation. Listed below is a combination of common principles from a variety of sources.[6,14]

1. Evaluation is part of program design.
2. Not everything needs to be evaluated.
3. Methods follow purpose.
4. Outcomes and impacts are segregated into two categories depending on how they were achieved: (a) a concentrated and sustained program, or (b) single event or smorgasbord effort.
5. Stakeholder satisfaction and perception of program quality and benefits are essential to long-term success and the ability to transfer the program elsewhere.
6. Participants come to the program with some knowledge and experience.
7. Program participants are not a homogeneous group.
8. Specific factors related to program quality should be considered in an evaluation.

In 1981 the Joint Committee on Standards for Educational Evaluation[15] issued *Standards for Evaluations of Educational Programs, Projects, and Materials*. The standards reflect a consensus about the principles that partnerships must adhere to in an evaluation and provide a set of criteria by which the quality of an evaluation — and subsequently the results — can be judged. The committee identified 30 standards grouped into four attributes of a good evaluation: as utility, feasibility, propriety, and accuracy standards.

1. The utility standards are intended to ensure that an evaluation will serve the practical information needs of given audiences.
2. The feasibility standards are intended to ensure that an evaluation will be realistic, prudent, diplomatic, and frugal.
3. The propriety standards are intended to ensure that an evaluation will be conducted legally, ethically, and with due regard for the welfare of those involved in the evaluation, as well as those affected by its results.
4. The accuracy standards are intended to ensure that an evaluation will reveal and convey technically adequate information about the features of the object being studied that determines its worth or merit.

Just as an implementation strategy provides the framework for the implementation effort, a strategy needs to be developed for the evaluation component. There are several ways to develop an evaluation plan, and each of these ways needs to address several steps for the evaluation to have a chance of successful completion. A successful evaluation is when the project meets the evaluation criterion — is water quality improving? If the criterion is not met, then a weakness in the management effort can be identified and the implementation approach can be improved to achieve

Evaluation

water-quality goals, or the criterion can be determined to be inappropriate and modified. The following outlines the general steps in developing an evaluation plan:

1. Manage the evaluation. Who should run it — a team, the steering committee, or a consultant? How should the evaluation responsibilities be formulated? How much should the evaluation cost? How should evaluations tasks be organized or scheduled? What kinds of problems can be expected?
2. Focus the evaluation. What will be evaluated? What is the purpose of the evaluation? Who will be affected by the evaluation, and who needs to be involved in the evaluation? What are the crucial evaluation questions?
3. Design the evaluation. What are the alternative ways to design an evaluation? What does the design include? What is the structure of the evaluation?
4. Collect information to support the evaluation. What kinds of information need to be collected? What procedures need to be used in the data collection effort? How much information is necessary?
5. Analyzie and interpret data. How will the data be managed? What are the QA/QC procedures for data management? How will the data be analyzed? How will the results of the data analysis be interpreted?
6. Report. Who should get the report? What should the content of the report be? What is the appropriate structure of the report? What else needs to be done to make the report useful to stakeholders? What is the reporting schedule?

9.7 EVALUATION REPORTING

Evaluation reports are not meant to be general documents but rather, documents that provide information to decision makers and implementation teams on what is working, where changes are needed, and whether or not the project is moving in the right direction. The report needs to be credible.

Reporting evaluation results is an important component aspect of the evaluation process. Reporting information consists of four aspects: findings, interpretations, judgments, and recommendations. Findings are the facts and empirical results of the evaluations. Findings must be supported by sufficient data. Interpretations are the explanations offered about the findings, speculations about the interrelationships, causes, reasons for the findings, and meanings given to the data. Judgments are values brought to bear on the data. Recommendations are suggested courses of action to funders, steering and planning committees, staff, and others about how to improve the watershed management effort based on the findings, interpretations, and judgments. Evaluations should try to provide explanations about the success and failures of the activities, rather than just report on accomplishments. It is important for evaluations to explain failures rather than generate excuses for them. If possible, the evaluation should document changes in behavior or environmental improvements rather than just "bean count."

Shepard[16] lists several common myths of project reporting:

1. Complex analysis and big words impress people.
2. One report is enough.
3. Describing limitations weakens the report.
4. Impacts are obvious.
5. Audiences knows why they received the report.
6. Everything should be reported.

9.8 SUMMARY

Evaluations have a dark side. Think of your last performance evaluation. Harold Williams[17] points out, "From an early age, we have learned that if evaluation has point, it is one that pricks. Evaluation means testing, and the point of testing is to catch mistakes."

Several general observations can be made about evaluating watershed management efforts. A strong link must exist between the evaluation effort and the communication plan. The communication plan will detail how to communicate with the public and stakeholders. It will provide the "nuts and bolts" for tailoring the results, summary, and analysis to the right audience. Once the work has been done and you know the value of the watershed management effort, spreading the word and putting the results to work are important steps to follow. Applicability of any of the following individual observations will depend on the purpose of the watershed management effort.

1. The partnership has the responsibility to determine the evaluation aspects to be included in the watershed management effort.
2. Evaluation formulation should begin when the watershed management effort begins and be part of the planning and implementation processes. After-the-fact evaluations usually focus heavily on what was accomplished. They seldom do an adequate job of evaluating how the effort progressed from the beginning to the end, due to the lack of a baseline.
3. To be useful in decision making, evaluations must be timely and based on valid information.
4. Evaluation information that truly reflects the target audience is needed for decision-making purposes. Evaluations that focus on the average conditions and average operations are not useful in decision making since they tend to mask the issues related to the target audience with more issues from the public at large.
5. Hard data generally make steering committees more comfortable when used to make project decisions. However, hard data are usually more expensive and time-consuming to acquire. Hard data take analysis to make it useful.
6. Evaluation results must be presented in easily understandable terms.
7. For an evaluation to be useful, it must be accepted by the steering committee and the overall partnership. Acceptance of evaluations is more likely when the partnership is involved throughout the evaluation process.

For more information on evaluations, see *Evaluator's Handbook* by Joan Herman, Lyn Lyons Moore, and Coral Taylor Fitz-Gibbon, 1987.

Program Evaluation: A Practitioner's Guide for Trainers and Educators by R.O. Brinkerhoff, D.M. Brethower, T. Hluchyj, and J.R. Nowakowski, 1983.

For RESULTREACH, contact the The Rensselaerville Institute, Rensselaerville, NY.

REFERENCES

1. Kondolf, G.M. and Micheli, E.R., Evaluating stream restoration projects, *Environ. Manage.*, 19(1), 1–15, 1995.
2. Holmes, N., Post Project Appraisals of Conservation Enhancement of Flood Defense Works, Research and Development Report 285/1/A, National River Authority, Ready, U.K., 1991.
3. USEPA, Evaluation of Experimental Rural Clean Water Program, EPA-841-R-93-005, USEPA, Office of Water, Washington, D.C., 1993.
4. Davenport, T.E. and Kelly, M., Water Resource Data for Highland Silver Lake Monitoring and Evaluation Project Madison County, Illinois Phase IV, IEPA/WPC/86-001, Illinois EPA, Division of Water Pollution Control, Springfield, IL, 1986.
5. Hermann, J.L., Morris, L.L., and Fitz-Gibbons, C.T., *Evaluator's Handbook*, Sage Publications, Newbury Park, CA, 1987.
6. Wisconsin Department of Natural Resources (WiDNR), Wisconsin Department of Agriculture, Trade, and Consumer Protection, Dane County Land Conservation Department and Marathon County Land Conservation Department, Nonpoint Source Control Plan for the Lower Big Rib River Priority Watershed Project, WT-539-00, WiDNR, Bureau of Water Resources Management, Nonpoint Source and Land Management Section, Madison, WI, 2000.
7. USDA–Natural Resources Conservation Service (NRCS), Toledo Harbor Pilot Project, final report, USDA–NRCS, Columbus, OH, 1998.
8. Jordan III, W.R., Assessing restoration in four dimensions, *Volunteer Monitor*, 11(1), Spring 1999.
9. Davenport, T.E. and Kaynor, S., Watershed management works: the Lake Le-Aqua-Na project, *Land and Water, The Magazine of Natural Resource Management and Restoration*, 42(2), 25–27, 1998.
10. Leeds, C.F. et al., *Ohio Water Quality Projects Evaluation*, Ohio State University, Columbus, OH, 1995.
11. Beech, R. and Drake, A., *Designing an Effective Communication Program: A Blueprint for Success*, National Network for Environmental Management Studies Program, USEPA, Chicago, IL, 1992.
12. Franke, R.H. and Kaul, J.D., The Hawthorne experiments: first statistical interpretation, *Amer. Sociol. Rev.*, 43, 623–643, 1978.
13. Terrene Institute, *Clean Water in Your Watershed: A Citizens' Guide to Watershed Protection*, Terrene Institute, Alexandria, VA, 1993.
14. Shepard, R.L., Locally led conservation: what it means, how it works, and the role of conservation professionals, *Proc. SWCS Annual Conf.*, Myrtle Beach, SC, 2001.
15. Joint Committee on Standards for Educational Evaluation, *Standards for Evaluations of Educational Programs, Projects, and Materials*, McGraw-Hill, New York, 1981.

16. Shepard, R.L., Creating a Report, the Human Dimensions of Watershed Management, an Interactive Workshop, 1999 International Symposium on Society and Resource Management, Brisbane, Australia, 1999.
17. Williams, H.S., Learning versus evaluation, *Innovating*, 1(4), 1991.

10 Monitoring

It is what we think we know already that often prevents us from learning.

— Claude Bernard

10.1 INTRODUCTION

While the terms "monitoring" and "assessment" are familiar to people working on watershed management efforts, the words are often used interchangeably and inappropriately. To avoid confusion, we use the following definitions for this discussion:

1. Monitoring consists of collecting data on many characteristics of the waterbody and its watershed according to specific quality assurance and control protocols (see appendix section for more information on quality assurance and quality control).
2. Assessment consists of the evaluation of data and information to describe the condition and status of a resource and to make judgments on that condition.

The assessment and problem identification phase documents additional data needs, and the planning phase sets priorities. Some watershed projects tend first to focus on satisfying data gaps with extensive monitoring and modeling efforts, at the expense of addressing known protection needs. Since different teams are responsible for implementation and monitoring, both types of activities can occur simultaneously. Today's society believes that more data are always needed and that chemical concentration data are the best. Neither of these beliefs is necessarily appropriate. They need to be determined by comparing project needs with available data. By selecting and applying suitable methods for evaluating the data, the partnership can prevent the "data–rich, information-poor syndrome."[1] The monitoring component identifies how much and what type of short- and long-term monitoring needs to take place to meet the partnership's goals. Water-quality monitoring is essential for determining project results and evaluating the effectiveness of management activities.

The most fundamental step in developing a monitoring component is defining the effort's purposes. Monitoring goals are usually broad statements. The goals of any monitoring program should support:

1. Production of water quality and landscape information that the partnership and stakeholders need to make environmentally sound decisions.

2. Improvement in communication and knowledge about water-quality and landscape issues on a watershed basis.
3. Resolution and prevention of conflicts over environmental impacts through positive cooperation and information sharing.
4. Empowerment of citizens' groups performing the monitoring to work cooperatively with partnerships to ensure appropriate land-use decisions are made for water-quality management purposes.

Getting to Know Your Local Watershed[2] outlines the important features to be included in the monitoring efforts and delineates the major categories of natural feature uses and social trends.

The lack of effective monitoring limited the value of the model implementation program, some RCWP projects, and, based on available information, the USDA Hydrologic Unit Area, Demonstration, and Management Systems Evaluation Areas water-quality projects.[3] Water-quality monitoring and data analysis provide the basis for a direct assessment of existing conditions in a waterbody. Monitoring has to be more than a one-time measurement, which provides only a "snapshot" of conditions and is then insufficient to establish a baseline condition or to document the impact of management. Designing a monitoring component that supports the goals of the watershed management plan must include selecting sampling variables, developing a sampling strategy, locating monitoring stations, selecting data analysis techniques, and determining the length of the monitoring efforts and the overall level of effort to be invested.

10.2 WATERSHED MONITORING

Establishing a monitoring plan is a key element in the adaptive management of a watershed. A comprehensive watershed management monitoring component must include landscape and water resource monitoring. The watershed-level monitoring effort will enable the steering committee to measure progress toward meeting administrative and environmental goals and provide crucial information for guiding management decisions through successive iterations of the adaptive management cycle.

Good planning and communication need to occur before the beginning of any monitoring and evaluation efforts. The most important step to take when developing any monitoring effort is to answer the following questions: why, what, where, when, and how. Failure to answer these questions may result in the collection of data and information that may not serve the purpose for which it is needed.

The goal of data collection should be to provide decision makers with answers — what are the problems in the watershed, what are the causes of the problem, are the proposed actions going to lessen the problems? Decision makers must have confidence in the data to use them effectively. Monitoring requirements need to be defined by the data users (steering committee) and not the data collectors.

Development and implementation of watershed monitoring efforts that support narrowly defined objectives increase the likelihood that the monitoring results will be relevant and useful. A thorough understanding of the water-quality problem, monitoring objectives, and expected results will help the steering committee make informed adaptive management decisions while overseeing the implementation of the watershed management effort. The most defensible measure of a watershed management effort performance is a well-designed and implemented monitoring effort that statistically examines the relationship between the project's activities and changes in water quality. The monitoring program results should be able to check if assumptions about direct, indirect, and cumulative effects of management activities on resources are valid. To use data effectively and persuasively, the partnership's leadership must understand the reasons for collecting specific types of information and the methodology used to gather it. These reasons and methods should be explained in the watershed management plan, in a simple and straightforward manner understandable to the nonscientific layperson. Adequate and effective management activity and water-quality monitoring for the watershed management effort are required to document progress toward water-quality goals, assess existing management approaches, determine needs for future management, maintain the interest of stakeholders and funders, and ensure the partnership's credibility. In addition to supporting adaptive management, monitoring and assessment need to be conducted for several reasons. Relating these various reasons to the watershed management process helps define the timing and intensity of activities.

- Assessment phase. Gather and analyze information and data for assessment and problem identification purposes.
- Planning phase. Fill data gaps, refine the overall baseline, and set the stage for short-term effectiveness monitoring efforts.
- Implementation phase. Determine how much progress has been made and provide evidence that good management is making a difference.
- Phase four. Evaluate the implementation efforts and determine the impact and the need to modify the existing management approach.

Carefully defining and documenting the water-quality problem is one of the most important steps in watershed management. This allows the partnership to target management activities to specific problems. If the available information and data are insufficient to identify the problems adequately, additional information is needed. An effective approach to overcome this deficiency is to implement a problem identification monitoring and assessment effort. The effort may have to last four to six months to refine existing data and information; in some cases, it may last throughout the project to identify further problems. Problem identification monitoring uses a site-specific plan to identify pollution sources and impacts during base-flow and storm conditions. In order to be effective, problem identification monitoring needs to focus on the timeframes of greatest pollutant loading and when the impairments are present. The level of problem assessment depends on available information and data, the nature of the water resource's impairment, the diversity and complexity of pollutant sources, the hydrologic transport system, potential uses of the data in the TMDL process, and the watershed size. Accurate and complete problem assessment is instrumental to achieving water-quality goals. An evaluation of the problem and land use up-gradient from the water resource provides much of the information needed to specify the monitoring effort's objectives.

Baseline studies are required to provide more information on a site, to develop management goals, and to refine the monitoring component. This is often conducted during the planning phase and can be considered the initial phase of the monitoring effort. Baseline information is a very useful data set on existing conditions against which performance of the management approach can be evaluated. Monitoring during the implementation phase is done primarily to ensure the management plan is correctly carried out. Close communication is needed between the monitoring team and the implementation team to evaluate the impacts of the implementation efforts and to ensure necessary midcourse corrections are identified and proposed to the steering committee.

The overall scale of the monitoring effort has two components — temporal and geographic scales. The temporal scale is the amount of time required to accomplish the monitoring efforts. The geographic scale can be a very small parcel to a large river basin. The temporal and geographic scales, like the monitoring efforts design and levels of intensity, are primarily determined by the watershed management plan's goals and objectives. The West Branch Delaware River Model Implementation Project (New York) focused on barnyard controls to reduce phosphorus loading to the Cannonsville Reservoir. This intensive effort led to negligible reductions in phosphorus loads to the Cannonsville Reservoir. The follow-up evaluation found that phosphorus loadings from barnyards were less than the phosphorus in runoff from fields receiving manure in the winter. If an annual (temporal) pollutant budget that covered not only phosphorus generation but also disposal (geographic) were developed, the implementation efforts would also emphasize land application of manure rather than just barnyard management.[4]

The monitoring effort's objective needs to be related to the water-quality problem and the overall management objective. A substantial amount of time may be necessary to specify monitoring objectives, but this initial effort will improve the long-term efficiency of the monitoring effort. Identification of watershed monitoring component constraints for each objective should address financial, staffing, and temporal elements. Clear and detailed information should be obtained on the timeframe within which management decisions need to be made, the amounts and types of data that must be collected, the level of effort required to collect the necessary data, and the equipment and personnel needed to conduct the monitoring.

Annual pollutant budgets are useful decision tools to determine variables and frequency of monitoring and expected information from load monitoring from various sources. Chapter 5 uses White Clay Lake to highlight the need to create both pollutant and water budgets to support development of watershed-level pollutant control strategies. The Johnson-Sauk Trail Lake, Henry County, IL, is an example of the need to document the impact of pollutant sources, including *in situ* contaminants on an annual pollutant budget. A Federal Clean Lakes Phase 1 Study[5] documented that internal regeneration of phosphorus was the dominant source of phosphorus to the lake. The annual internal regeneration of phosphorus from bottom sediment was 75% of the dissolved phase and 66% of the total phase. The phosphorous budget emphasizes the need for the management approach to focus on in-lake techniques. The adopted management approach focused on removing phosphorous sources and binding phosphorous to the bottom sediment through destratification, selective harvesting and removal of weeds, and control of algal blooms using chelated copper sulfate application followed by potassium permanganate application. Another example is the St. Albans Bay RCWP (Vermont), a 10-yr land treatment program targeted at the suspected major phosphorus sources; it was unsuccessful in improving bay water quality. An intensive monitoring effort documented two other sources of phosphorus — the recycling of phosphorus from bay sediments and the elevated levels of phosphorus in agricultural soils — were the reasons for no improvement in bay water quality.[6]

BOX 10.1

Monitoring the loading rate is a very useful measure for evaluating current conditions and trends in pollutant loading, developing TMDLs, and evaluating the effect of the land treatment. The loading rate, or mass of pollutant exported per unit time, is a basic measurement for eutrophication studies and pollutant budgets. Loading rates are directly comparable to one another, but they can vary significantly from year to year. There are two ways to determine loading rates: direct water-quality monitoring and surrogate estimates.

The three major tasks for determining pollutant loads by water-quality monitoring are:

1. Measuring water discharge (ft^3/sec)
2. Measuring pollutant concentration (mg/L)
3. Calculating pollutant loads (multiplying discharge times concentration over a year)

Estimating 80 to 90% of the annual load to be delivered during 10% of the year is common.[7] This timeframe corresponds to periods with high pollutant fluxes. Depending on the pollutant evaluated, fluxes during snow melt and storm events are often many times greater than those during low-flow conditions. Thus, the monitoring program must be designed to account for periods of highest pollutant flux. In a review of evaluative studies of loading approaches, Richards[7] highlights several points concerning selected methodologies. Averaging methods generally have a bias that decreases as the number of samples decreases. Regression approaches can perform well if the relationship between flow and pollutant concentration is well defined, linear throughout the range of flows, and constant throughout the water year. Ratio approaches performed better than regression approaches, and both are superior to the averaging approaches. If a project is to use one of these approaches, the teams need to be aware of these issues. Greater detail and examples regarding these approaches can be found in Richards.[7]

10.3 TYPES OF MONITORING

Three types of monitoring are possible: physical, chemical, and biological. A number of guides and manuals provide the specific techniques and methodologies; they are listed at the end of the chapter. The discussion here focuses on concepts and issues related to the project's management. Biological assessment and evaluation provide the most realistic indicator of the resource's ecological condition.

For monitoring the state of biological variables, the length of the life cycle needs to be considered when determining the sampling interval. Sampling should be repeated within a year for systems where temporal variability is estimated for the year or season and for measures of its variability. Consider also the seasonal changes and the life cycle for biota. The minimum sampling frequency may be two times the length of the life cycle for most biota.

Chemical monitoring is used to ascertain compliance with water-quality standards (usually concentration-based) and loading estimates for development, implementation, and evaluation for TMDLs. For chemical monitoring the type of water sample limits the data's usability. The type of sampling selected determines the population to be sampled, the statistical approach to be used on the data, and the choice of sampling equipment. Richards[7] makes several general observations regarding sample type and estimating loads. Accuracy and precision increase with increased frequency of sampling. Grab (point or instantaneous) samples are insufficient to determine loads unless concentrations are correlated to discharge, which needs to be measured continuously. Stratified random sampling with most samples taken during periods of high flow can provide increased precision for a given number of samples. Time–weight composite samples are not sufficient for load estimation because they do not reflect changes in discharge and concentration during the period over which the samples are composited. Flow-weighted composite samples are well suited to load estimation.

10.4 MONITORING PURPOSES

Working with the watershed management process' phases and specific "whys," the team can categorize the monitoring purposes (Table 10.1). Traditional categories are problem investigation, condition monitoring, compliance monitoring, and effectiveness monitoring. Monitoring only one type of change may not show progress as quickly as a more diverse monitoring strategy will. Each type of monitoring answers different questions; most efforts have a combination of monitoring approaches integrated on a watershed basis. For example, the Lower Big Rib River Project has three

TABLE 10.1
Monitoring Purpose by Watershed Management Phase and Reason

Watershed Management Phase	Why	Monitoring Purpose
Problem identification/assessment	Establish baseline; determine trends in water quality	Condition and problem investigation monitoring
Plan development	Establish goals and objectives; fill data gaps	Condition and problem investigation monitoring
Implementation	Track progress; use trend analysis to detect directional changes	Compliance and condition monitoring
Evaluation	Determine trend and impact; need for midcourse corrections	Condition and problem investigation monitoring

monitoring efforts: volunteer monitoring, "signs of success" monitoring, and master monitoring sites. The combination of monitoring efforts supports the progress evaluation of the nonpoint source program, short-term monitoring effort designed to see if management made a difference, and long-term trend analysis.[8] It is important to note that early progress backed by actual environmental data strengthens implementation efforts and makes the project more attractive to funders. Designing a constant and consistent monitoring effort is the key to demonstrating progress. Table 10.1 relates the watershed management phases with the purposes of monitoring. Depending on the monitoring design, some monitoring purposes such as condition monitoring are applicable for all phases of the management process.

Condition, baseline, or ambient monitoring addresses the question, "What is the status of the water resource?" or "Is the waterbody's quality sufficient to meet water-quality standards?" In addition to describing existing conditions, this type of monitoring can be used in long-term trend assessments. Condition monitoring requires a comparison of observed water-quality conditions with desired water quality, expressed as a reference condition, criterion, or standard. According to Coffey,[9] sufficient baseline data are required for impact assessment because baseline or historical monitoring is fundamental to analyze the problem and system function, and to document variability. Baseline monitoring during premanagement actions is usually required to detect a trend or impact or to show causality-related management. Depending on the monitoring design, up to 2 years of premanagement monitoring and 2 to 5 years of postimplementation monitoring are typically needed for small watershed (≤7000 acres) studies. Wisconsin's Master Monitoring Site program involves 11 years of monitoring, including 2 years of premanagement practice installation, 5 years of monitoring during implementation, 2 years of monitoring during the response period, and 2 final years of monitoring during the postpractice installation phase.[10] Less time should be needed for edge of field studies, when hydrologic variability is known to be less than typical for larger watersheds and subwatershed systems, or when a paired watershed design is used. Condition monitoring is usually done at a state or regional level through a network. Multiple years of consistent condition monitoring are needed for trend monitoring. A time series must be obtained to document changes in water quality due to management actions including the implementation of management practices. Measurement using comparable, consistent methods should be taken either at regularly timed intervals or for specified periods for a sufficient length of time. The components of a time series are both deterministic and random. Deterministic components change in a predictable manner and are assumed or known without error. The random component is measured with error and consists of unexplained factors that hinder the trend's detection. To detect a trend, the random component and complex deterministic factors such as cycles, and the dependence of one observation on the next (serial correlation), must be considered. For trend detection, the monitoring objective must be translated into a testable statistical hypothesis to provide structure to the monitoring design. This type of monitoring is used to establish trends in water quality and is appropriate for the problem identification and assessment phase (phase 1), tracking implementation of management activities (phase 3), and evaluating long-term impact and progress (phase 4).

Problem investigation monitoring is usually more intensive and discrete than condition monitoring. In some situations where neither the source nor cause of an impairment is known or the implementation activities are not having the expected effect, a special monitoring effort is needed to determine what they are. This type of monitoring does not just identify a problem, but provides a more thorough description of the problem and its sources. This type of monitoring is often done to follow up questions regarding a specific source or problem. Problem investigation monitoring can be at the watershed assessment level or problem-specific such as the identification of critical areas. This type of monitoring usually fills data gaps (phases 1 and 2) or refines a problem further (phase 3). Other problem investigation monitoring efforts include responding to emergencies, performing research, and calibrating/validating water-quality models.

Compliance, implementation, or regulatory monitoring is used to assess whether specific performance standards or requirements are met. This type of monitoring will not provide a link to stream-water quality. Compliance monitoring is required by the NPDES permit system and most state/tribal water quality programs for industrial and municipal water dischargers. This type of compliance monitoring tracks the status of point source implementation and operation (phase 3). Landowners who have concerns about regulatory enforcement need to understand the monitoring efforts. Landowners in the Taylor Creek/Nubbin Slough watershed were appreciative of information produced with the synoptic surveillance component of the overall RCWP watershed monitoring program to identify and alleviate trouble spots.[11] Compliance monitoring of nonpoint source pollution controls is land-based and tracks management practice implementation and their operation/maintenance. Land treatment and land-use monitoring (tracking) are needed to quantify the location, extent, and quality of land treatment. Quantitative monitoring of implemented land treatment allows documentation of trends in land treatment. The methods of reporting and quantifying land treatment and land use should be consistent over the implementation period and related to the watershed management plan. An uncomplicated method should be developed that limits the amount of paperwork required to document implementation while still allowing a minimum level of tracking. Table 10.2 lists monitoring trends by monitoring type.

TABLE 10.2
Monitoring Frequency, Duration, Data Analysis Intensity, and Scale by Monitoring Type

Type of Monitoring	Frequency of Sampling	Duration of Sampling	Intensity of Data Analysis	Scale of Monitoring
Condition	Low	Trend: long Baseline: short	Trend: low to moderate Baseline: low	Subwatershed: watershed
Problem	High	Short	Medium to high	Subwatershed: watershed
Compliance	Variable	Duration of effort	Low to medium	Variable
Effectiveness	Medium to high	Short to long	Medium	Variable

Evaluation or effectiveness monitoring is designed to measure the actual impact of resource management decisions, such as implementation of nutrient management practices and resulting impacts on nutrient levels in the waterbody. The monitoring takes place in specific locations and provides a measure of whether and to what extent the problems were addressed. Effectiveness monitoring involves monitoring before and after implementation activities have occurred. Five common designs are used in effectiveness monitoring. Each design relies on a different level and intensity of monitoring and provides different answers at various level of confidence. Table 10.3 shows when monitoring occurs and the questions various effectiveness monitoring designs answer. "Signs of success" monitoring under the Wisconsin Nonpoint Source Program is an excellent example of pre- and postdesign.[10]

Water-quality monitoring traditionally begins with the implementation of land management practices. In these circumstances it is difficult to establish a positive correlation between management practice implementation and water-quality impacts. The ability to combine water-quality and land treatment data on appropriate spatial and temporal scales with pre- and postmanagement data is essential to determine if the trends in water quality match the mechanistic prediction in trends. Monitoring teams will have difficulty detecting a statistically significant impact if the design of the monitoring program does not incorporate a control watershed or if it is not monitored successfully. Sources of variability include climate, weather, watershed characteristics, and human activities. Variability may be in daily, seasonal, or year-to-year patterns or have some random component. Measuring and accounting for sources of variability increase monitoring sensitivity and reduce minimum detectable change. Explanatory variables should be selected because they measure factors in the ecosystem thought to affect the primary variable(s) of concern. Incorporation of explanatory variables helps isolate water-quality changes due to land treatment. For example, the Nansemond-Chuckatuck Project (Virginia) targeted activities on reducing nutrients, sediment, and pesticides to downstream reservoirs. While the management approach focused on treating land to reduce downstream nutrient loading, groundwater high in phosphorous was pumped into project area reservoirs when the reservoirs were low, thus confounding detection of phosphorous trends.[3]

Typically the stronger the conclusions the partnership wants to draw from the information collected, the more information it will need to collect. What are the tradeoffs among the various designs for collecting original information and data in terms of their depth, cost, and skill level of the collector? The level of effort needed to support data management and analysis needs to be identified. Few partnerships appreciate the enormous effort required to analyze data. Given the random nature of rainfall events and the varied character of runoff, it is often necessary to commit extensive resources to provide statistically valid results for rainfall–runoff-related variables. Mosteller and Tukey[12] identify four conditions needed to show causality or cause-and-effect: association, consistency, responsiveness, and a mechanism. Association is shown by demonstrating a relationship between two variables (e.g., a correlation between intensity of management and the apparent reduction in pollutant loadings). Consistency can be confirmed by observation only and implies the relationship does not change in different populations (e.g., management action was implemented in several areas and pollutant loading was reduced, depending

TABLE 10.3
Selected Characteristics of Effectiveness Monitoring

Effectiveness of Monitoring Design	When Monitoring Occurs	Questions Being Answered	Supports Adaptive Management
Postdesign	When the management effort has been completed	What is the condition after the management effort has been completed?	NA
Pre- and postdesgin	Before and after management effort	What is the change in condition before and after management activity?	NA
Pre- and postdesign with multiple samples in between	Before, during, and after management	What is the change in condition and direction of trend in water quality?	Yes
Pre- and postdesign with control group	Before and after for matched or paired watersheds	What is the change in condition, and is there a correlation with management?	Yes
Comprehensive design	Multiple before, during, and after in various designs with controls and matched/paired watersheds	What is the change in condition, and did the proposed management activities cause it?	Yes

on the effect of treatment, in each case). Responsiveness is shown in an experiment when a treatment is performed with a corresponding change in a variable. A mechanism is a plausible step-by-step explanation of how the management action could cause the observed change. For example, conservation tillage reduced the edge-of-field losses of sediment, thereby removing a known fraction of pollutants from runoff to a stream. The result was decreased suspended sediment concentration in the water column.

Failure to observe improvement may mean that the problem was not correctly documented, such as in the case of the Highland Silver Lake RCWP; management actions were not directed properly, as noted in the Rock Creek RCWP; the type of management actions implemented were improper, as in the Haines Creek Project. The Conestoga Headwaters Project (Pennsylvania) suffered because the strength and intensity of the management activities were inadequate, the monitoring effort was not sensitive enough to detect change, or more time was needed for the effects to be documented. Traditionally, a midcourse evaluation, if conducted appropriately, provides an opportunity for modifications in management approaches or monitoring design. Reevaluation of management activities and refining the land treatment efforts are possibilities if new information is gained to refine the partnership's approach to addressing the water-quality problems and their sources. The value of this approach was documented with the Garvin Brook RCWP. This project maintained enough flexibility to allow for programmatic adjustments throughout its duration. Groundwater monitoring and hydrologic analysis prompted adjustments, which included expansion of the project area to include all the groundwater recharge area for the Garvin Brook watershed and a shift in emphasis from surface-water to groundwater protection management practices and monitoring.[13]

Monitoring by objectives may require a different approach for each. Monitoring to evaluate current conditions should focus on critical variables expected to respond to management activities. For violations of water-quality standards, the choice of variable is specified by the standard or criterion of concern. To assess ecological integrity, partnerships need to monitor a set of variables that show how an ecosystem compares to a control or reference condition that has a composition, structure, and function essentially unimpaired by human activities.[14] For trend detection, the response variable and explanatory variable must be carefully selected to show management effect and account for changes in system variability. Monitoring type is related to the objectives and goals of the watershed management plans. Table 10.4 presents some examples of this relationship and responsibilities.

Monitoring objectives must be narrowly and clearly defined to address a specific problem at an appropriate level of detail. Spatial and temporal information related to the problem is essential for implementing a successful monitoring program. The monitoring objective specifies, where appropriate, the primary variables, the degree of causality or other relationship, and the anticipated result of the management action. For example,

> To evaluate current conditions in Ethan Creek by analyzing the ecological integrity and suitability of the creek as a water supply

TABLE 10.4
Examples of Linking Watershed Use/Issue, Goal, and Responsibility by Monitoring Type

Use/Issue	Goal/Objective	Monitoring Type	Responsibility
Drinking well water	Reduce ambient nitrate concentrations to 10 mg/L by 2005	Condition: groundwater	State water quality agency
Fishery	Increase fishery by 5% over the next 10 years	Condition: fish and macroinvertebrate populations	State natural resource agency
Swimming	Reduce fecal concentrations to allow swimming	Condition: surface water and problem investigation of fecal sources	Health Departments, State water quality agency

To document the water-quality problems in Lake Emily by identifying specific pollutant constituents, their magnitude, sources, and impacts on the designated uses of Lake Emily

To detect trends in nitrogen concentrations in Craddock Creek due to land application of animal waste

To evaluate the impact of a municipal wastewater treatment plant upgrade on algal blooms in Pebbles Lake

Tracking water quality and informing the public on progress increase the likelihood for attaining the desired level of land management implementation. The steering committee must decide on the level and intensity of the tracking effort. All changes in land use should be monitored, not just management controls, to help distinguish those water-quality changes associated with controls from naturally occurring changes. Land-use changes that affect water quality include conversions and intensity of use. The evaluation of the Maumee River Pilot Sedimentation Reduction Project documented that conversion of land to cropland would result in the management approach's failing to meet its soil erosion reduction goals.[15] Accounting for all major sources of variability in water-quality and land management data increases the ability to isolate true water-quality trends due to management activities. Correlation of water-quality changes and land management changes, by itself, is not sufficient to infer causal relationships. Other factors unrelated to management activities may cause changes in water quality, such as changes in land use, rainfall patterns, etc. Factoring in these explanatory variables can account for water-quality change due to climatic variables over time. All sources of variability in land management and water quality should be taken into account. For example, in the Prairie Rose Lake RCWP project, lake drawdown and rotenone treatment of carp impede the detection of lake water-quality trends, further complicating the linkage between water quality to land management. The problems of stream-bank erosion in the Rock Creek Watershed continues to mask in-stream benefits from installation of management practices in the watershed.[3]

Management activities are expected to affect physical, chemical, and biological variables; therefore, an integrated monitoring approach that accounts for ecosystem components is desirable. Physical monitoring has two components: a watershed survey of landscape conditions and habitat assessment. The landscape-level monitoring must be able to document management actions and change in landscape condition. The monitoring program must be consistent with the watershed management plan's goals. Thus, if the plan outlines program activities, management practice installation, and environmental goals, ideally the monitoring component provides the data to measure progress on all three goals.

Management activity and land-use monitoring are used to track where and when management practices are implemented. Tracking should also document how well the implemented practices adhere to the watershed management plan. The purpose is to track the treatment's strength in time and space. When tracking management activities, select explanatory variables that will accurately reflect the desired management effect. Management impacts can be expected in three areas: source area (acres treated), delivery area between source and receiving ecosystem (number of acres treated by activity), and direct ecosystem effects. A major challenge in attempting to relate implementation of pollutant controls to water-quality changes is determining the appropriate land management attributes to track. Land treatment and land-use monitoring should be related directly to pollutants or impacts monitored at the site.[9]

10.5 SCALE OF MONITORING

Three spatial scales exist for watershed monitoring: field or practice, subwatershed, and watershed outlet. Criteria for selecting the spatial scale are the monitoring objective, location and intensity of management, type of waterbody system, funding, and availability of sampling equipment.

10.5.1 FIELD

Edge-of-field monitoring is ideal for land management demonstration and pilot studies but is not recommended as an approach for documenting the overall effectiveness of a watershed management project. Monitoring at this level usually provides information on field-level conditions representative of the physiography and land-use activities impacting water quality and uses. Field-scale monitoring is always needed to document the effect of innovative land management practices.

10.5.2 SUBWATERSHED

Monitoring at the subwatershed level by taking samples close to pollutant sources and the land management activities can be useful for observing the aggregate effect of implementation on a group of parcels (fields) or several management areas. Subwatershed monitoring networks measure the aggregate effects of new management activities and existing management runoff as it enters an up-gradient tributary or receiving waterbody. Subwatershed monitoring can also be used for targeting critical areas. Subwatershed can be monitored in order to develop a precipitation-runoff model to estimate its relationship with hydrologic response and pollutant export. At this scale, partnerships should be able to see evidence that better land management does make a difference. Monitoring of land treatment and land use on a subwatershed scale was performed in Idaho, Vermont, Illinois, Oregon, Pennsylvania, South Dakota, Florida, and Nebraska so that land management could be correlated with water-quality monitoring from the same subwatershed.[3]

10.5.3 WATERSHED

Monitoring at the watershed scale is appropriate for assessing project impacts and pollutant load using a single station. Depending on station arrangement, both sub-watershed and watershed outlet studies are very useful for water and pollutant budget determinations. Monitoring at the watershed outlet is the least sensitive of the spatial scales for detecting management effect. While monitoring at the watershed scale is characterized as outlet monitoring, it is important for the partnership to monitor the drainage throughout the watershed in order to check progress and document impact. The RCWP projects in Michigan, Vermont, Idaho, Utah, Virginia, Pennsylvania, and Florida found that subwatershed monitoring the overall project area is more effective than monitoring only at the watershed outlet. Sensitivity of the monitoring program decreases with increased basin area and decreased treatment extent, or both. In addition, nontreatment effects such as hydrologic variability and nonhomogeneous land uses increase minimum detection change (MDC) levels at the watershed level. For RCWP projects, Spooner et al.[16] developed a method to calculate the MDC for water-quality parameters (fecal coliforms, total phosphorus, and suspended sediment) to help guide individual projects' monitoring efforts. Reckhow et al.[17] developed a method to determine the number of sampling events required to detect a statistically significant change of a given magnitude and range of errors.

10.6 VOLUNTEER MONITORING

Water-quality monitoring, which involves the collection of water samples to measure chemical, biological, and physical indicators of waterbody health, has become a common and well-established domain of citizen volunteer organization. Citizen volunteer monitoring of water, habitat, and ecological quality has evolved as a main component of scientific data collection. However, citizen monitoring cannot replace professionally collected water-quality data, but it does add a critical complementary dimension to data collection efforts. The benefits include supplementing the data collected through government collection programs and educating a new set of stakeholders. The greatest benefit of volunteer monitoring is that it promotes development of a citizenry that is not only educated about water quality issues but also is personally involved and committed within a watershed. While this advocacy is one of greatest benefits of citizen monitoring, it can put agencies in a dilemma since they usually do not have the resources to respond immediately to citizens' requests.

This, in turn, causes frustration and disappointment for citizens. A second concern is associated with agency treatment of the data volunteers collect. Unless the volunteers' data are collected and analyzed by the same procedures as those used by the water monitoring agency, the volunteer data set is usually treated differently because of the procedures and methods used. Further, the acceptance of volunteer data by a water-monitoring agency does not necessarily give the data more credibility. The quality of the volunteer monitoring effort is what gives it credibility.

Agencies are recognizing that, with proper training, local volunteer groups can be relied upon to inventory resources and perform monitoring in their watersheds. In some instances volunteer groups are forming partnerships with agencies to conduct ongoing, systematic monitoring of lakes, streams, and riparian resources. These volunteer partnerships are a key to reversing the problem of data "gaps" or "voids" for most of our nation's waters. Volunteer or citizen-based data collection programs also increase the general public's (1) involvement in watershed management and (2) awareness of water issues and the importance of government data collection programs.

It is important to clearly define the goals and objectives of a citizen monitoring effort and ensure all participants are in agreement with them prior to program initiation. It is just as valuable to a watershed partnership to support volunteer monitoring for the purpose of learning about the watershed's wetlands, streams, or lakes as it is to identify water-quality problems and track trends. Shoreline and stream surveys work well as an educational and screening tool and can be used at the outset of any volunteer monitoring effort. In terms of sustainable, long-term watershed management, the educational aspects are the most important to build the necessary public support. The GLOBE program is an excellent example of this type of effort. GLOBE is a worldwide hands-on primary and secondary school-based science and education program. GLOBE provides the opportunity for students to learn by taking scientifically valid measurements, reporting the data through the Internet, and collaborating with scientists and other GLOBE students around the world.

10.7 DEVELOPMENT OF A MONITORING EFFORT

Watershed monitoring programs have evolved from routine collection of grab samples into complex levels of environmental monitoring technology. The RCWP was instrumental in this evolution. The RCWP's 22 projects served as living laboratories for monitoring and evaluation efforts in the U.S. Many lessons were learned throughout the program's evolution. One lesson is to spend time up front designing the monitoring effort and implementing what was designed. Unfortunately for the Tillamook Bay RCWP project, the monitoring strategy changed during the project, and it was further handicapped by inadequate and limited background data. Based on this experience a sampling strategy should be established and followed throughout a project's life. At a minimum for dairy coastal water-quality management efforts, the parameter set must include data on fecal coliform bacteria, bay salinity, river flow, and rainfall. Moore and others[18] were not able to document statistically valid trends because of data limitations.

Deciding when to monitor each pollutant and associated explanatory variables is difficult. The overall issue of when to monitor has several parts: concerning initiation (when to start), timing (whether to monitor randomly, regularly, or in response to a condition), frequency (how often), duration (length of the effort), and consistency. A number of processes are available for a partnership to use to develop a monitoring effort. They range from the NRCS's[19] 12-step process for monitoring program design to Sanders et al.[20] and the state of Minnesota's five-step process for monitoring system design. The five-step process is straightforward and provides the flexibility needed to develop a watershed-specific monitoring effort that can be used to support the watershed management process. The five steps are:

1. Evaluating information expectations and needs. The first step in the effort is determining what information is needed to evaluate the water quality and natural resource concerns of a watershed. This determination includes identifying the concerns, evaluating previous studies, identifying information goals and needs, and developing monitoring objectives that support the information goals. Objectives and budget dictate the level of monitoring detail. Levels differ primarily in terms of skill, intensity, time, resources, questions being answered, and equipment necessary. An aspect of time most overlooked is the timeframe within which decision makers require monitoring information.
2. Establishing design criteria. The second step is developing design criteria to outline the procedures that will be used to evaluate a watershed's water quality and natural resource concerns based on the goals and objectives developed in the watershed management planning process.
3. Designing the monitoring system. The third step is designing the actual network. Based on watershed characteristics, management approaches, and problems, a monitoring design that involves the selection of parameters to be analyzed, sampling site, and sampling frequency is completed in the third step. The monitoring stations can be categorized by waterbody type, location, scale, and purpose: tributary, main stem stream, wetland, lake, groundwater, and receiving waterbody. A mixture of station types (depending on the cost and situation) would be useful in documenting problems. The TMDL Federal Advisory Committee[21] identified the following factors affecting how much data are necessary to support a TMDL:
    The extent of follow-up monitoring for which the TMDL calls
    The potential impact on the environment
    The potential impact on sources
    Data needs of models and other tools necessary to develop an approvable and scientifically defensible TMDL
4. Developing operating plans and procedures for the monitoring system. The next step in the design process involves the development of operating plans and procedures for sampling, sample handling, laboratory analysis, quality assurance/quality control (QA/QC), and data management. During this step data management, logistics, record keeping, and backup procedures are established. Maintaining field QA/QC is essential to the success of monitoring programs and to ensure data quality.[11] It is well known that

TMDL development has been inhibited by inadequate data collection, incompatible data from different sources, and failure to follow proper analytic techniques in data collection.[21]

5. Developing information reporting procedures. The last step in the process is the development of information reporting procedures linked to the partnership's communication plan. This step provides the link to the steering committee's decision-making process and the public.

A major factor in designing a monitoring effort is cost. Traditionally, planning committees see monitoring costs in terms of a tradeoff with implementation activities rather than as an equally important factor in the overall success of the watershed management effort. The cost of monitoring is relatively low compared with the benefits. Under the RCWP, monitoring costs about 16% of the total program; the Clean Lakes Program's monitoring costs were in the range in of 5 to 10%. Monitoring is not a tradeoff activity.[3] Monitoring costs are generally governed by four factors:

1. The number of monitoring stations/sampling sites. Any monitoring effort conducted on a watershed scale must include a decision about the watershed's size. The number of stations needed to assess conditions and to track improvements over time is largely a function of watershed size, complexity, management approach, and issues addressed.
2. The number of sampling events. The number of sampling events monitored is largely a function of project duration, the type of problem addressed, and the necessity to obtain statistically valid results. The time between samples or sampling interval and the number of sampling events or years of monitoring are key elements of the sampling design.
3. The types of parameters monitored. The parameters monitored are related to the issues addressed and to level of confidence needed in the answer. If water chemistry is balanced with biological and physical indicators, cost can be significantly reduced while still providing the data needed to assess the efficacy of the watershed management program. In order to provide a more holistic picture of current and historical conditions, biological and physical parameters should be included.[22]
4. Availability of funds. The total amount of funding is a limiting factor for all monitoring and data analysis efforts.

10.8 REPORTING

It is particularly critical to ensure the steering committee and other interested stakeholders (1) know the type of information the monitoring component will produce, and (2) have realistic expectations concerning what can be done with the information. Several key steps ensure realistic expectations are placed on the monitoring component.[23] The following are the most applicable steps about the watershed management process.

1. Review the administrative structure and procedures the partnership developed to define the information expectations of the steering committee.

2. Review the ability of the monitoring effort to supply the information.
3. Formulate an information expectations report for the monitoring effort. This needs to be linked with the social capacity building effort.
4. Present the information expectations report to all users of the information. This needs to be linked with the outreach strategy.

When developing an information expectations report, select the presentation of results based on the audience reviewing the information and the purpose of the monitoring component. At a minimum, the data analysis group should prepare example report formats for the steering committee to approve. The purpose should always be to present clear and accurate information that is not subject to misinterpretation. This effort needs to be defined in the partnership's communication plan.

The best single source of general information on watershed monitoring programs is available through the watershed academy web site: http://www.epa.gov/owow/watershed/wacademy.

Additional resource materials include the following:

USEPA, *Monitoring Guidance for Determining the Effectiveness of Nonpoint Source Controls,* EPA 841-B-96–004, Office of Water, USEPA, Washington, D.C., 1997.

NRCS, *National Handbook of Water Quality Monitoring,* Part 600, National Water Quality Handbook, 1996.

Ponce, S.L., *Water Quality Monitoring Programs,* WSDG-TP-00002, USDA Forest Service, Fort Collins, CO, 1980.

USGS, *National Handbook of Recommended Methods for Water-Data Acquisition,* Office of Water Data Coordination, USGS, U.S. Dept. of Interior, Reston, VA, 1977.

For tracking nonpoint source controls, see:

USEPA, *Techniques for Tracking, Evaluating, and Reporting the Implementation of Nonpoint Source Control Measures – Agriculture,* Office of Water, USEPA, Washington, D.C., 1996.

USEPA, *Techniques for Tracking, Evaluating, and Reporting the Implementation of Nonpoint Source Control Measures – Forestry,* Office of Water, USEPA, Washington, D.C., 1996.

For more information on the GLOBE Program, see www.globe.gov.

REFERENCES

1. Ward, R.C., Loftis, J.C., and McBride, G.B., The data rich but information poor syndrome in water quality monitoring, *Environ. Manage.,* 10, 291, 1986.
2. Conservation Technology Information Center (CTIC), *Getting to Know Your Local Watershed, a Guide for Watershed Partnerships,* CTIC, West Lafayette, IN, 1995, p. 8.

3. USEPA, Evaluation of Experimental Rural Clean Water Program, EPA-841-R-93-005, USEPA, Washington, D.C., 1993.
4. Brown, M.P. and Rafferty, M., *A Historical Plan for Research and Monitoring of the West Branch Delaware River Model Implementation Program*, New York State Dept. of Environ. Conservation, WM-912, 1978, p. 35.
5. Kothandaraman, V. and Evans, R.L., *Diagnostic-Feasibility Study of Johnson Sauk Trail Lake*, Illinois State Water Survey (ISWS) Contract Report 312, ISWS, Urbana, 1983.
6. Meals, D., *Lake Champlain Basin Agricultural Watersheds Section 319 National Monitoring Program Project*, final report, May 1994–Nov 2000, Vermont Dept. of Environ. Conservation, Waterbury, 2001.
7. Richards, R.P., *Estimation of Pollutant Loads in Rivers and Streams: A Guidance Document for NPS Programs*, Water Quality Laboratory, Heidelberg College, Tiffin, 1979.
8. Wisconsin Department of Natural Resources (WiDNR), Wisconsin Department of Agriculture, Trade, and Consumer Protection, Dane County Land Conservation Department, and Marathon County Land Conservation Department, *Nonpoint Source Control Plan for the Lower Big Rib River Priority Watershed Project*, WT-539-00, WiDNR, Madison, 2000.
9. Coffey, S.W., Spooner, J., and Smolen, M.D., *The Nonpoint Source Manager's Guide to Water Quality and Land Treatment Monitoring*, NCSU Water Quality Group, Raleigh, NC, 1995.
10. WiDNR, Wisconsin Department of Agriculture, Trade, and Consumer Protection, Dane County Land Conservation Department, and Columbia County Land Conservation Department, *Nonpoint Source Control Plan for the Lake Mendota Priority Watershed Project*, WT-536-00-REV, WiDNR, Madison, 2000.
11. Osking, K. and Gunsalus, B., The evolution of RCWP water quality monitoring networks in the Taylor Creek/Nubbin Slough and Lower Kissimmee River Basin, in *Proc. the Rural Clean Water Program Symposium, 10 Years of Controlling Agricultural Nonpoint Source Pollution: The RCWP Experience*, EPA/625/R-92/006, USEPA, Washington, D.C., 1993.
12. Mosteller, F. and Tukey, J.W., *Data Analysis and Regression: A Second Course in Statistics*, Addison-Wesley, Reading, MA, 1977.
13. Wall, D. et al., Understanding the groundwater system: the Garvin Brook experience, in *Proc. The Rural Clean Water Program Symposium, 10 Years of Controlling Agricultural Nonpoint Source Pollution: The RCWP Experience*, EPA/625/R-92/006, USEPA, Washington, D.C., 1993, pp. 59–70.
14. Karr, J.R. et al., Assessing Biological Integrity in Running Waters, a Method and Its Rationale, Illinois Natural History Survey, Champaign, IL, special publication 5, 1986.
15. USDA-Natural Resources Conservation Service (NRCS), *Toledo Harbor Pilot Project;* final report, USDA-NRCS, Columbus, 1998.
16. Spooner, J. et al., Increasing the sensitivity of nonpoint source control monitoring programs, in *Proc. Monitoring, Modeling, and Mediating Water Quality Symp.*, May 1987, American Water Resources Association, Middleburg, VA., 1987, pp. 242–257.
17. Reckhow, K.H. et al., Case study: detecting of trends and sampling study evaluations, in Wedepohl, R.E. et al., Eds., *Monitoring Lake and Reservoir Restoration*, EPA 440/4–90–007, prepared by North American Lake Management Society for USEPA, Washington, D.C., 1989.
18. Moore, J.A., Pederson, R., and Worledge, J., Keeping bacteria out of the Bay — the Tillamook experience, in *Proc. The Rural Clean Water Program Symposium, 10 Years of Controlling Agricultural Nonpoint Source Pollution: The RCWP Experience*, EPA/625/R-92/006, USEPA, Washington, D.C., 1993, pp. 71–76.

19. USDA, *National Handbook of Water Quality Monitoring*, Part 600, USDA, Washington, D.C., 1996.
20. Sanders, T.G. et al., *Design of Networks for Monitoring Water Quality*, Water Resources Publications, Littleton, CO, 1983.
21. USEPA, Report of the Federal Advisory Committee on the Total Maximum Daily Load (TMDL) Program, EPA-100-R-98-006, USEPA, Washington, D.C., 1998.
22. Yoder, C., Answering some concerns about biological criteria based on experiences in Ohio, *Proc. Water Quality Standards for the 21st Century*, 1990, pp. 95–104.
23. Ward, R.C., Loftis, J.C., and McBride, G.B., *Design of Water Quality Monitoring Systems*, Van Nostrand Reinhold Company, NY, 1990.

11 Models

The sciences do not try and explain, they hardly even try to interpret, they mainly make models. By a model is meant a mathematical construct which, with the addition of certain verbal interpretations, describes observed phenomena.

— John Von Neumann

11.1 INTRODUCTION

The use of computer models in watershed management as an evaluative and predictive tool is expanding. Models allow partnerships to ask "what if" questions. Watershed planning involves the consideration of how different strategies for activities, land uses, and system designs might affect and improve the health of a watershed. In an atmosphere of constrained budget, limited data, and insufficient time, many planning committees are turning to models to aid in the evaluation of impacts associated with various alternative management strategies. The increased availability and accuracy of models now make them a more valuable tool in quantifying water-quality goals. For example, the development of TMDLs often requires the use of watershed loading models to evaluate the effects of land uses and practices on pollutant loading to a waterbody. A model can provide a frame of reference for considering this problem within a watershed context.

This discussion on modeling was developed to assist partnerships evaluate the options available for using computer modeling in their efforts. Most watershed partnerships make a distinction among water-quality models between (1) environmental models focusing primarily on the chemical and physical processes in a waterbody, and (2) economic models focusing on the costs of pollution control strategies. Both types of models are important in watershed management. In actual practice, since most of the cost models available are oriented toward unique, local situations, their transferability to other locations is more difficult so they are not used. Therefore, the model discussion here focuses on the environmental type.

A partnership's interest in a model needs to be balanced between two factors: the cost, in terms of time and finances, of obtaining the information, and the value of the information generated through the model application. This chapter introduces the idea of watershed computer modeling and, in general, the types of models available, specifically surface-water-quality models. It also covers program and technical criteria that can be used to select models to address specific watershed needs, and factors to consider in model use and results analysis. The alternate approach to using technical and program criteria is to first determine the resources available to invest in a modeling effort, select the model that most nearly approaches the limit of available resources, and then determine what objectives can be achieved using the selected model and its results. The latter approach is not recommended.

The term "model" generally denotes the physical or mathematical representation of a physical system. Critical in the selection of an appropriate modeling approach is identifying the functions the model is expected to perform and determining how well the model needs to perform these functions. TetraTech, Inc.[1] completed an evaluation of ten different models that could be used to estimate the effectiveness of potential management practices to meet performance standards associated with urban nonpoint source control requirements for total suspended solids (TSS) loadings. TetraTech, Inc. worked with its clients to develop five criteria to rate the applicability of the various models. Two critical criteria were used in the final selection: ability to be used as a screening tool, and ability to evaluate the effectiveness of various management practices in reducing TSS loadings. Two of the ten models reviewed were recommended for possible use. Defining modeling needs in terms of purpose, complexity, and methodology can serve as an initial screening tool to eliminate certain classes of models from consideration. Then other criteria such as availability of data, schedule, and cost can be used in the final selection. In all modeling efforts, it is important to recognize the limits of knowledge. No model can be more accurate than having actual data and measurements. Only validated models should be used to evaluate environmental improvements in a watershed project.

Model output must have value toward improved problem definition and decision making. Model-generated outputs are considered an effective tool for visualizing the potential results during the planning process. Models are only tools to help partnerships and technical staff in carrying out their responsibilities, but the danger exists that managers may interpret model output as the absolute truth, ignoring the influences of data quality, modeling assumptions, and the randomness of the natural physical processes. Even under the best circumstances, modeling results should be considered as estimates, since the model itself is only an approximation of the real watershed system.

11.2 MODEL CATEGORY

Models can be categorized in a number of ways. Different models offer different capabilities including both the types of processes that they can address (e.g., quantity versus quality) as well as what level of detail (e.g., annual loadings of pollutants versus hourly simulations). This variety of model capabilities allows a watershed partnership to select and use a model that addresses the specific concerns within a watershed at an appropriate level of detail. Models can be used not only to evaluate alternative management approaches, but also to assess current and potential future problems in the watershed.

The most common categorization is between physical and mathematical models. In a physical model a physical structure is built, such as a hydraulic model, in which water is allowed to flow within the model and measurements are taken. A mathematical model can be defined as a formulation of simplification that expresses the essential features of a physical system or process in mathematical terms.[2] The mathematical models can be further subdivided in a number of ways. According to Fleming,[3] three broad categories of mathematical models exist: deterministic, statistical (correlation or regression), and optimum search (optimization). The majority of the models used in watershed management are deterministic, with varying levels of physical basis or a combination of deterministic and statistical.

Huber and Lickinson[4] utilize model purposes (screening, planning, design, and operational) as a categorization technique for runoff models. Screening models are preliminary desktop procedures to provide an initial estimate of stream flows and pollutant loads. Planning models are used for overall watershed assessments as well as for initial estimates of the effectiveness and costs of various management strategies. Design models are oriented toward the detailed simulation of individual storm events and are effective tools for determining a least-cost control strategy for water-quality and quantity problems. Operational models are used in producing actual management decisions during storm events.

The USEPA[5] identifies three categories of models for estimating pollutant loading: overview, midrange, and detailed. The defining characteristics of models are the degree to which processes are simplified and the time scale that is used for analysis and display of output information. These loading model categories are based on complexity, operation, time step, and simulation technique. The level of detail required, type of outputs, user-friendliness of the model, hardware and software requirements, and the need to calibrate the model for specific application vary by category. The overview category models are the simplest and easiest to apply and the detailed are the most difficult. For a comparative review of nutrient loading models see USEPA,[6] for sediment see USEPA,[7] and for pathogen models see USEPA.[8]

In addition to the three categories of models defined by the USEPA, a number of models go beyond estimating pollutant loadings and sources and attempt to predict water-quality responses to changes in pollutant loading. Such models can be helpful in establishing project goals and objectives as well as in evaluating alternative management strategies. Also, such models can be used to evaluate the cumulative impacts of land development on various receiving waters. BASINS (Better Assessment Science Integrating Point and Nonpoint Sources) is an integrated modeling system for performing watershed and water-quality studies.[9] BASINS includes assessment tools, spatial data, and watershed and water-quality modeling components, with GIS providing the integrating framework. The Lost River in West Virginia was listed on the state 303(d) list due to elevated levels of fecal coliform from a variety of sources. A model that had an ability to integrate point and nonpoint source simulation, as well as an ability to assess in-stream water-quality responses, was needed to complete the necessary TMDL. BASINS was selected since it met these criteria. The modeling results indicated that there was no need to reduce point source discharges and that with the following load allocations: 38.34% reduction from pasture land, 12.8% from forest, and 37.75% from cropland, the Lost River could meet water-quality standards. With these reduction targets as goals, an implementation plan that promoted manure storage and application guidelines, crop and pasture management, and wildlife management was developed.[10]

11.3 MODEL APPLICATION

Most basic overview models provide a rapid means of identifying critical areas with minimal effort and data requirements. Annual or seasonal budgets can be generated using these models. Models such as the simple method, USLE/MUSLE, and USGS regression are considered overview or simplified models that require no calibration. This type of model relies on generalized sources of information and has low requirements for site-specific data. Overview models provide only rough estimates of pollutant loading and have limited predictive capability, and their results should be field-verified. The watershed treatment model (WTM)[11] is a simple tool for rapid assessment of various urban watershed treatment options. WTM assesses two broad categories of pollutant sources: primary land uses and secondary sources. WTM's primary land uses are the typical land-use categories that are in most geographic information system (GIS) land-use data layers. Secondary sources are pollutant sources, dispersed throughout the watershed, which are either wastewater-derived or a human activity such as construction. Information on secondary sources is not readily available based on land-use information and must be estimated or generated through surveys. WTM incorporates many simplifying assumptions that allow planners to assess various programs and sources not typically tracked in midrange or detailed models.

Midrange models use a management-level approach to assessing pollutant sources and transport in watersheds and responsiveness to objectives and actions appropriate to watershed management planning; models such AGNPS, GWLF, and SWAT fall in this category. These models can be used to evaluate pollution sources and impacts over a broad geographic range and can assist in determining which areas within a watershed should be targeted for control efforts. These models are good for working with seasonal or storm event problems and require limited or no calibration. Such models are usually simple and intended to identify problem areas within large drainage basins and make preliminary, qualitative evaluations of management practice alternatives. For example, another approach to estimating nonpoint source loads relies on GIS and event mean concentrations (EMC). Much of the pollution entering Corpus Christi Bay is from nonpoint sources. Each land use in the 75-mi drainage basin is assigned an EMC, and this is correlated with geo-located mean annual runoff estimates using GIS. Point source pollutant loads were combined with the estimated nonpoint source loads in the spatial database. The combined loads were routed to receiving water, using a digital terrain model, and used to calculate an equilibrium concentration in the bay system. A significant benefit of developing the model was the ability to assess implementation strategies. This comparison was done by assigning management practice values to the various land uses and estimating the projected loads.[12] Depending on the application and the required accuracy level, a midrange model should be validated against storm events and daily loads prior to use in decision making.

Midrange models are helpful when the implementation team must decide whether a specified system of land management practices will meet a goal. For example, when planning to reduce the off-site impacts of soil erosion, an implementation team can evaluate various designs of hydrologic retention and sedimentation ponds by making use of erosion predictions from models for design calculations. The designer could use the model to predict the effect of anticipated future land-use changes on sediment delivery to the practice and adjust the design accordingly. Modeling alone is often inadequate to evaluate the success of management practice systems in reducing nonpoint source loading in agricultural settings because there are mixed land uses that change annually (rotation) and these land uses have different loading rates.[13] An additional issue is that most loading models fail to address stream channel and bank dynamics adequately, including the impact of management practice systems on these factors. A computer model was used to show watershed planners in Brooks Creek (Newaygo County, Michigan) what their watershed would look like if current development trends continued. This build-out analysis was then compared to another scenario that assumed an ordinance was developed to protect all stream corridors in the watershed. Based on the results of the model, planners in the Brooks Creek watershed concluded that they needed to develop an ordinance to protect the stream corridors in the watershed.[14] Midrange models have been used in the TMDL process to address nutrient issues.[6,7] The application of the generalized watershed loading functions (GWLF) model for nutrient TMDL development in the Tar-Pamilco Basin in North Carolina[6] and Lake Macatawa watershed[15] are examples of its use. Table 11.1 presents a summary of the modeling and data assumptions for TMDL process components.

Detailed models incorporate the current understanding of the physical watershed processes affecting pollution generation and transport. Models such as HSPF and

TABLE 11.1
TMDL Component and the Needed Modeling, Data, and Assumptions

TMDL Component	Modeling, Data, and Assumptions
Target identification	Develop numeric target for water-quality conditions; translate criterion to numeric loading capacity level
Deviation from target	Quantify the amount and timeframe of deviation between current/future loading levels and the loading capacity level
Source identification	Identify all sources or source categories; quantify the amount of load from sources, including natural background
Allocation of pollutant loads	Ensure that allocations will lead to attainment of water-quality standards
Implementation planning	Estimate the effectiveness of controls/management practices; determine that controls/management practices are sufficient to achieve the TMDL allocations; determine the likelihood of actual implementation of management strategies
Monitoring/evaluation	Assess whether the implementation of controls/management practices has occurred; evaluate the effectiveness of controls/management practices and whether they are meeting allocations; demonstrate attainment of water-quality standards

Source: USEPA, 1998, Report of the Federal Advisory Committee on the TMDL Program, 26 p.

SWMM are considered detailed models, and calibration is recommended for their use. Detailed models mainly use continuous simulation to predict flow and pollutant concentrations for a range of flow conditions. It must be understood, as with simulation of any natural systems, that mathematical models are only a rough approximation. The use of detailed models requires a considerable amount of time and resource expenditures. Detailed simulation models are extremely data-intensive. This type of model has also been used to determine the necessary components in a watershed management implementation effort. A GIS-based modeling approach was used in the Waukegan River Project to characterize the physical and hydrologic features of the watershed spatially.[16] GIS made it possible to construct high-resolution digital models of the watershed for environmental modeling utilizing AUTO-QI to evaluate alternate management practice application strategies. The AUTO-QI modeling indicated that with the application of traditional urban management practices, the water-quality problems in the Waukegan River would not be adequately addressed.[17] After an alternatives analysis, the Waukegan River project was then expanded to include aquatic restoration in order to restore beneficial uses and address the identified water-quality problems.

When a field-scale detailed model is used to estimate the effectiveness of various management practices in reducing nonpoint source pollution at its source within a large watershed, a number of representative fields should be selected for modeling due to the cost. Knisel[18] defines a field as a small area (3 [1.3 ha] to 75 [30.4 ha] acres) with relatively homogeneous soils under a single management practices that is small enough that rainfall variability is minimal. Davenport[19] utilizes a detailed field model — CREAMS (chemical, runoff, and erosion from agricultural management systems) to estimate the relative effectiveness of management practice systems in reducing sediment pollution in the Highland Silver Lake watershed project. Output from the CREAMS model allowed planners to look at the quantity of sediment, but also at the particle-size distribution of the sediment. This was valuable since the sediment pollution of Highland Silver Lake was related to the composition of the sediment rather than its volume. Creating a unitless index using modeling outputs allowed for the relative comparison of management practice systems in terms of pollution potential and alleviated the baggage associated with managing to "T" and farmers thinking they are protecting the environment by managing soil productivity. Utilizing a potential pollutant index generated from CREAMS modeling runs, landowners/operators were shown that no-till management had the greatest impact in reducing their potential off-field pollution associated with sediment. The potential pollutant index is a measure of the amount and fineness of sediment and is the product of sediment yield and enrichment ratio.

Chandler[20] makes an excellent point when comparing the variability of data typically utilized in detailed runoff simulation models. Field studies he reports on found that stormwater runoff pollutant concentrations can vary by an order of magnitude or more for comparable storm events at a given location, and average rainfall can vary by a factor of 2. Since these two major input parameters can vary to such a high degree, it would be inappropriate to expect any model to estimate annual loads to a greater degree of accuracy. Stallings et al.[21] recommend that modelers identify the most important input variables, concentrate on getting good estimates for those variables, and estimate the remaining variables as well as possible. In addition for urban models, the utility of the soil information must always be closely examined. Most modelers use the USDA county level soil surveys when developing their soil's information; for urban areas most soils are disturbed and no longer reflect the original soil characteristics. For most urban situations, the top layers of the existing soils were compacted during construction; the topsoil either was imported from another location or had been modified with amendments or contaminants, or both. In agricultural situations, Stallings and others[21] find field-scale model results can vary up to 100% when soil data are generalized within an individual field. The availability of the appropriate soil information at the level of detail required should be a selection criterion when looking at detailed field models.

Models have become essential for performing TMDL analyses at the watershed scale. Understanding that all models abstract from reality, modelers accept tradeoffs between model complexity, data requirements, transparency, and the quality of the model results. Modelers should make choices when selecting and using models based on technical assessments among tradeoffs and not model and data availability. Most stakeholders have little or no experience with models. Modelers need to spend the time to build knowledge in stakeholders so they can actively participate in the modeling effort. Local confidence in the model and the consequential TMDL results are essential to building support for implementation. One way to build this confidence is improved communication between modelers and stakeholders regarding model assumptions. Usually stakeholders are not aware of the tradeoffs inherent in the modeling process and therefore have high expectations for model accuracy. The accuracy associated with the results of the modeling studies depends on the specific model used, the accuracy of the input data, the characterization of the watershed being simulated, and the expertise/experience and resources available to the model user. In addition, while models can be used to provide information, they cannot determine which values should guide the TMDL development.

Most models are often more accurate in a relative sense than in an absolute sense. Some models are tools for a quantitative analysis of specific watershed problems or issues; they do not provide simple yes-or-no answers to watershed partnerships. Instead, they provide information about the expected response of the watershed to a given management proposal that must be analyzed and interpreted in a logical and consistent fashion in order to assist the watershed planning committee in making decisions. Clearly, the key in the use of any model is the analysis and interpretations of the model outputs. The output must have value toward improving problem definition and aiding decision making. There is always the danger that managers will interpret model output as truth, ignoring the influences of data quality, inherent model assumptions, and randomness of the watershed processes. The user should be aware of three terms associated with modeling: calibration, validation, and verification. Calibration is testing the model with known input and output used to adjust or estimate factors. A good calibration using bad data is a bad calibration. Validation is comparison of model results with an independent data set (without further adjustments). Verification is the examination of the numerical technique in the computer code to ascertain that it truly represents the conceptual model and that there are no inherent numerical problems. For calibration purposes there is an increase in the required level of effort to calibrate a model from an annual balance to storm event time step due to the level of detail information required. The priorities for calibration/validation procedures are hydrology (first), sediment (next), and water quality (last). Rules of thumb for calibration/validation accuracy vary by parameter; for example, hydrology (5–10%), erosion (10–35%), sediment transport (20–50%), and other water-quality pollutants (10–40%). The water-quality data set used in the calibration and validation process needs to be representative of the true distribution of water conditions in the watershed. Representative data sets must reflect concentration values covering the range of flows and land management conditions in the watershed. The land-use and management data need to be linked to the water-quality parameters simulated and to the pollutant sources and management practices to be implemented.

When looking for a model to predict receiving water quality for TMDL determinations, look for a model that is able to transport the pollutant through the system addressing the hydrologic and hydrodynamic regime of the water system; chemical, physical, and biological reactions within the system; and inputs and withdrawals from the system. The type of waterbody to be monitored determines the model dimensions; spatial dimensions of receiving water models are one-dimensional — longitudinal (rivers), two-dimensional in the vertical (lakes) and two-dimensional in the horizontal (sediment transport), and three-dimensional — estuaries. Models such as BATHTUB, QUAL2E, WASP, HSPF, and EFDC are receiving water models. Surface-water and groundwater models can be integrated to evaluate the effects of land-use changes on groundwater and surface-water resources. USDI-GS[22] worked with the city of Middleton and the Wisconsin Department of Natural Resources to correlate the effects of urbanization on stream flow and spring flow in the Pheasant Branch watershed. The spring system is an important water resource in the watershed and an important source of water for a wild rice community. The output of the surface-water model was coupled to the groundwater model input to simulate how urbanization affects the stormwater runoff and groundwater recharge. The study explored the potential benefits of various mitigation measures associated with urbanization. The city of Middleton used the results to impose additional requirements on a development in the recharge area of the springs and to be proactive in the location of future developments and the development of green space in the watershed.

11.4 MODEL SELECTION

According to Chandler,[20] the first step in any modeling process is to determine whether or not a model is required. Once determination has been made that a water-quality model is necessary, the choice of appropriate model must be made through a process of defining project needs and limits. Figure 11.1 depicts a decision process for the selection of the appropriate model for pollutant loading and TMDL development in cases where modeling was deemed necessary. The following list will aid in the model selection process.

1. Have a clear understanding of what the model is to accomplish in terms of the watershed effort.

Models

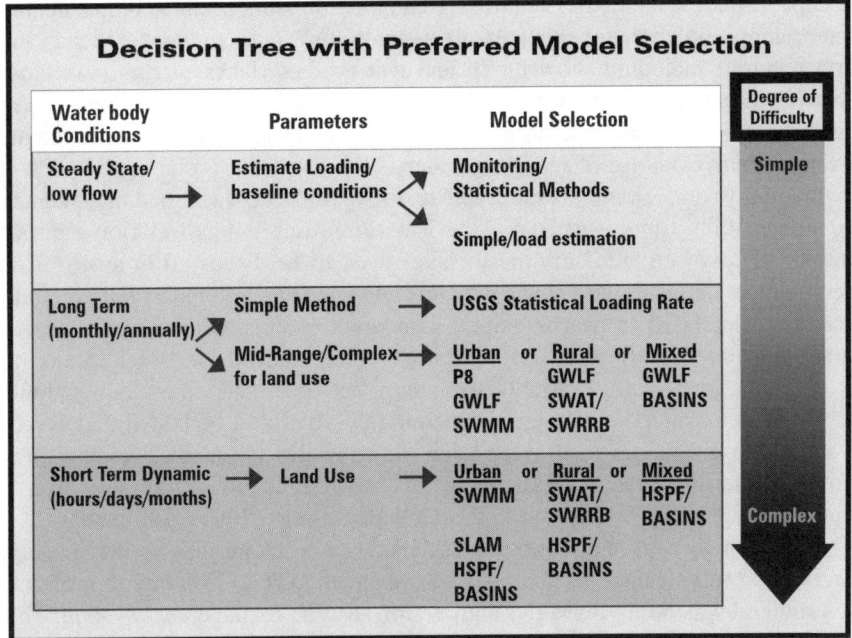

FIGURE 11.1 Decision tree with preferred model selection process. (Courtesy of CTIC, West Lafayette, IN.)

2. Establish the objectives for the modeling.
3. Determine the appropriate timeframe to be evaluated through the modeling effort. The appropriate timeframe is established by project objectives and the necessary technical requirements. For example, if the partnership wants to establish an annual pollutant loading goal, the model needs to be able to generate annual loading estimates under various management scenarios.
4. Determine the application parameters (area, pollutants, data inputs, and output variables, etc.).
5. Identify the input requirements. Are sufficient data (quality and quantity) available to allow a model to be calibrated and validated? If not, can the necessary data be collected through the project in a timely manner?
6. Use the simplest model that will satisfy the project needs within the resource constraints of the project.
7. What are the data requirements for both the model and problem being analyzed?
8. What computer hardware and staff are required?
9. What documentation is available?
10. How much will it cost?
11. With proper data, how accurately will the model represent the real world?

If it is determined that a detailed model is needed, the selection criteria need to include a link between the model outputs and the decision-making process. Then the capability of a detailed model to simulate the following needs to be considered: meteorology (precipitation, temperature), hydrology (stream flow, groundwater, surface runoff including snow melt), and waterbodies (lakes, streams, wetlands, estuaries, oceans); spatial (watershed, subwatersheds) and temporal (annual, seasonal, event-based, continuous) simulations; land-use categories; and environmental effects on beneficial use of receiving waters.

In order to understand the modeling results within the watershed management decision-making framework, the analysis assumptions and expectation accuracy must be clearly defined. Both these issues need to be discussed in terms of the specific model considered. How a model is designed and applied usually includes assumptions related to processes and data requirements. Model results should be supplemented with sensitivity and uncertainty analysis in order to analyze the potential "real-world" variability about model-predicted values. For example, the Macatawa Watershed Project[15] utilizes a combination of USLE and GWLF to estimate phosphorus loading to Lake Macatawa in order to develop a phosphorus reduction strategy. The Macatawa watershed has a drainage area of approximately 110,000 acres (175 mi^2) with almost 70% of it managed as cropland. For this project, modelers assumed for cropland that no management practices were currently on the land. However, the CTIC[23] reports that 65% of the cropland was being managed with a crop residue management system. Thus the amounts of soil erosion and phosphorus were overestimated by the modeling effort, creating a false baseline on which to base the phosphorus reduction strategy. Overestimating the severity or significance of certain problems can divert attention from those that actually deserve greater concern. Decision makers need to be aware of the assumptions, and their associated limitations, in order to understand the applicability of the modeling results. Consideration of uncertainty of the simulated results is as least as important as the results in any decision-making process. Because of the complexity of quantifying modeling uncertainty, modelers are encouraged to devise the best approach for the modeling application.[5]

Gordon[24] raises a number of issues that model users must consider:

1. Inaccurate models that misrepresent reality can distort decisions.
2. The lack of understanding and support by decision makers can hinder modeling efforts.
3. The model is meant to assist, not replace, planning.
4. A model's usefulness to planners is contingent upon its ability to simulate the impacts of realistic decisions.
5. Modeling can never replace monitoring. However, modeling can be used to guide monitoring efforts.

Reviewing these issues, the planning and technical committees will assist the partnership in selecting and appropriately applying any necessary models to assist in watershed management plan or TMDL development.

Models

11.5 MODEL DOCUMENTATION PLAN

When utilizing a model as part of a watershed management effort it is important that the modeling process be documented. The purpose of having the documentation is to provide a firm understanding of what the modeling effort represents to the public and planning committee. At a minimum, the model documentation plan should include the following:

1. Model name and version.
2. Source of model.
3. Purpose of model application.
4. Model assumptions (list or summarize). If any of the assumptions could limit usability of the results of the application, it must be listed and explained.
5. Data requirements and source of data sets.

11.6 GEOGRAPHIC INFORMATION SYSTEMS

GISs are "systems that integrate layers of spatially oriented information whether manually or automatically."[25] Figure 11.2 shows the integration of various data layers. Computerized overlay maps developed through GIS have popularized its use

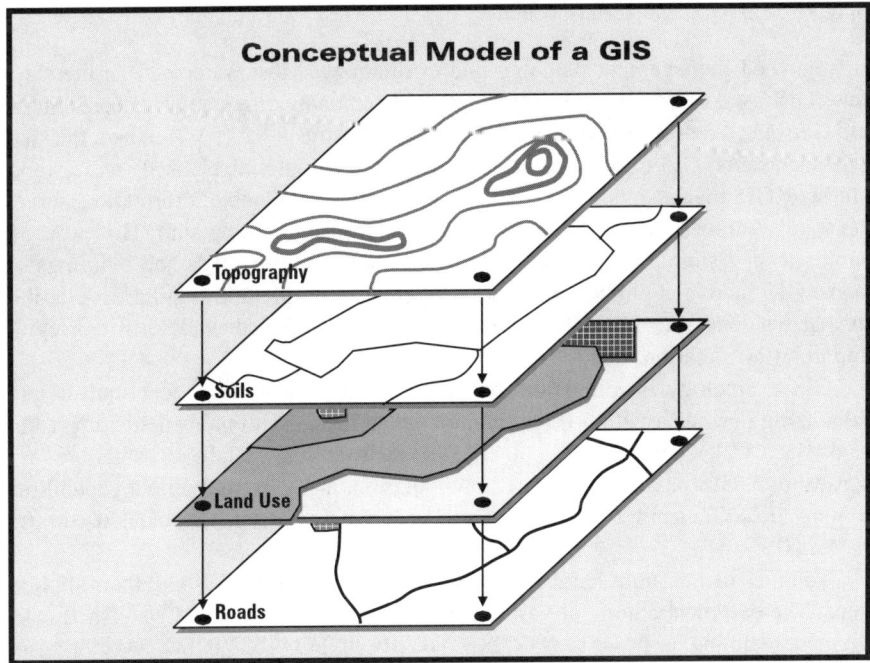

FIGURE 11.2 Conceptual model of a GIS. (Courtesy of CTIC, West Lafayette, IN.)

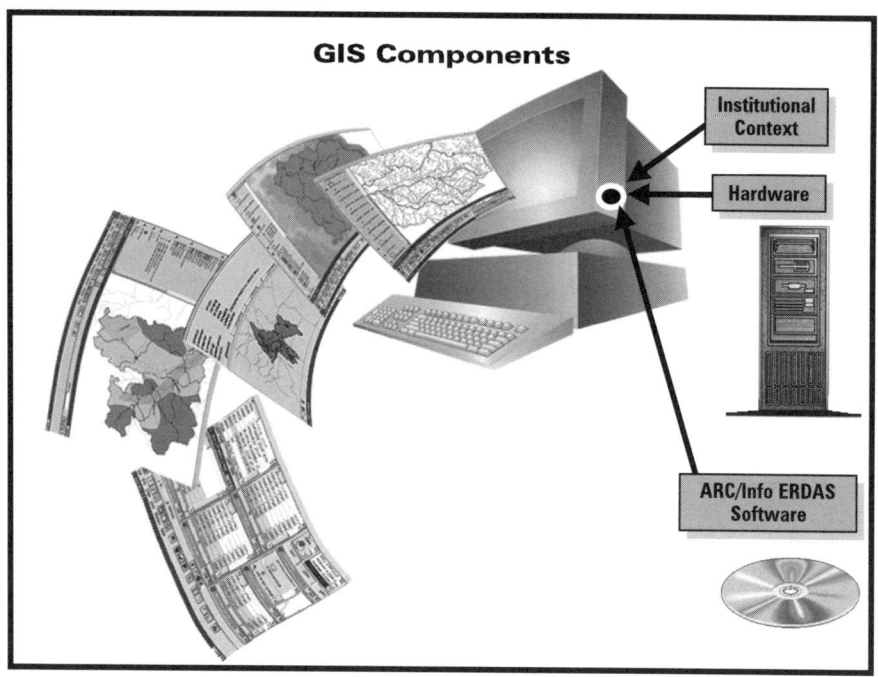

FIGURE 11.3 GIS components. (Courtesy of CTIC, West Lafayette, IN.)

in watershed management planning and evaluations. Most watershed partnerships view GIS as a powerful computer-based tool for analyzing spatial data. DeShazo and Garrigan[26] report that for the public in the Merrimack River Watershed, the state (Massachusetts) GIS maps were in many cases too complex to be easily understood. The state GIS maps contained a tremendous amount of valuable information, served the state's purposes, and were used by local land-use planning staff. However, the public needed simple information that they could relate to roads and buildings on maps. GIS maps for public use need to be kept as simple and familiar as possible so that residents can identify landmarks in their watershed while still conveying important information.

GIS technology integrates database operations such as statistical analysis and calculating new information from combined data layers with the benefits offered by maps. In a watershed project context GIS would include databases, software, and hardware. A GIS has input, storage, manipulation, analysis, and output capabilities (Figure 11.3). Table 11.2 highlights some of the major benefits associated with the use of GIS.[27]

Volumes of literature have been written about different GISs and their applications. The two most commonly used GISs for nonpoint source and watershed management planning purposes are GRASS and ARC/INFO. ARC/INFO, a vector-based GIS, is the predominant system in use today and is commercially supported by the Environmental Systems Research Institute. In vector-based format, information is stored as series of x-, y-coordinates or points. Vector-based systems are useful for

TABLE 11.2
Major Benefits Associated with GIS Use

Quicker access to information
Improved decision-making capabilities
Increased public awareness and acceptance
Increased analysis capabilities
Improved data management

Source: Griffin, C.B., *GIS Introduction for Public Agencies*, CTIC, West Lafayette, IN, 1995a.[27]

mapping discrete features such as watershed boundaries and drinking-water well locations, but are not as useful for mapping continuously varying features such as soil type. Geographic resources analysis support system (GRASS) is a raster-based GIS developed and maintained as a public domain system by the U.S. Army Corps of Engineers, Construction Engineering Research Laboratory.[28] A raster database is a collection of cells in a grid. Figure 11.4 shows the difference between geographic data represented in a vector and raster format.

Data quality is a major issue for GIS-generated outputs. Griffin[29] highlights the six components of data quality: lineage, positional accuracy, attribute accuracy, logical consistency, completeness, and temporal accuracy. These components are applicable to both source and operational errors. Planning committees, working with their GIS staff, must decide which data quality components are most important for their use since optimizing all six components could be cost-prohibitive. The nature of the questions asked should be used to help determine the data quality needed.

Griffin[27] reports four major uses of GIS in environmental problem solving. First, GIS-generated maps are a powerful tool for communicating spatial data and information. Second, GIS can be used to answer spatial questions concerning location of specific points and what is associated with that point. Third, GISs are used to manipulate data inputs to models and outputs from models. Fourth, environmental modeling can be performed in the GIS. GIS plays an important role in watershed management and nonpoint source control. For the land management component of typical watershed management projects, a GIS can be used to:

1. Prepare maps of watershed and subwatershed units.
2. Analyze the potential for pollutant contamination by subwatershed.
3. Prioritize watersheds for further nonpoint source planning and implementation efforts.
4. Identify and inventory potential sites for nonpoint source control installation.
5. Develop and evaluate pollutant control strategies, particularly land-use controls.

A GIS was used to estimate nonpoint source loading for the Corpus Christi Bay watershed. The approach was based on calculating the runoff from the individual

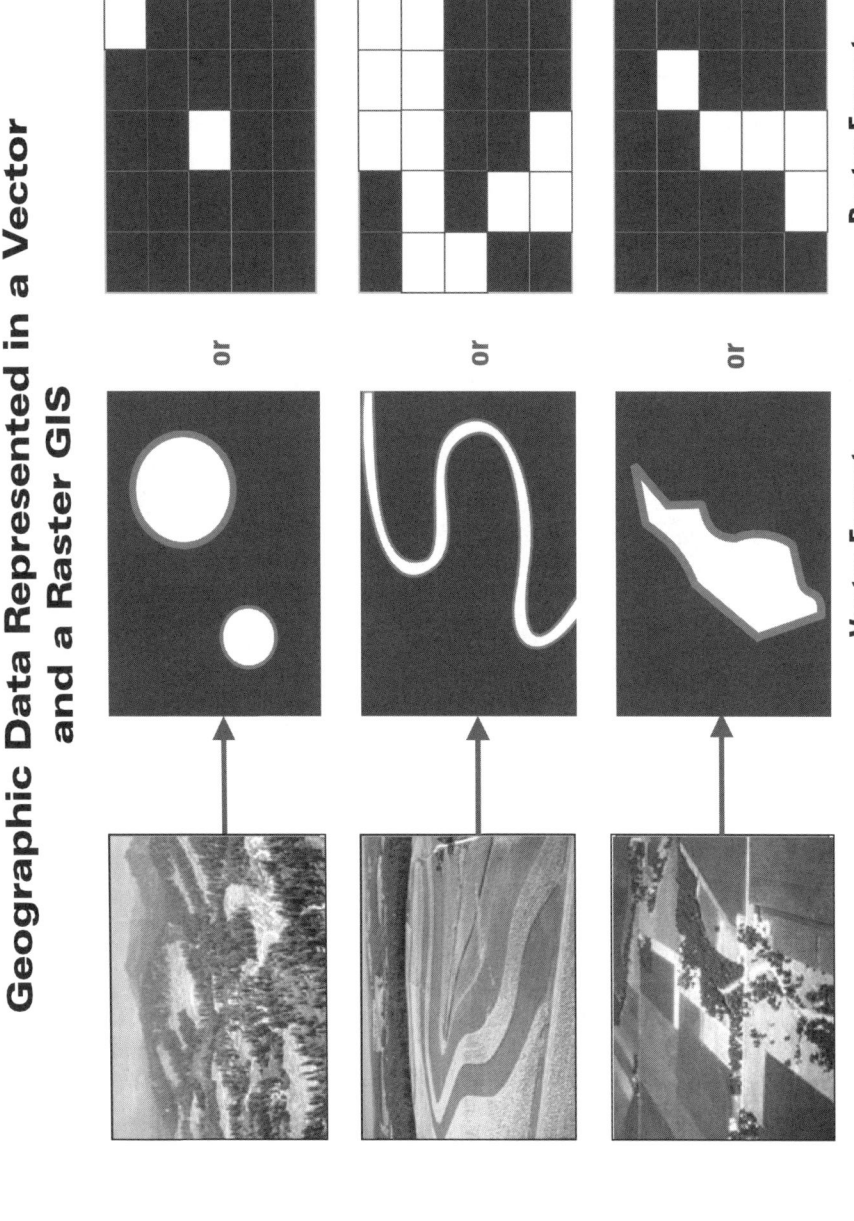

FIGURE 11.4 Geographic data represented in a vector and a raster GIS. (Courtesy of CTIC, West Lafayette, IN.)

land-use cells and then multiplying the runoff by the event mean concentration. The point sources were then assigned to a particular cell. GIS's ability to represent both point and nonpoint source loading on a watershed scale makes it an extremely useful tool in the development of TMDLs. Chen and others[30] report TMDL stakeholders like GIS maps for regional pollution loads and because they depict waterbody sections meeting water-quality standards in green and sections not meeting standards in red.

The utility of GIS application heavily depends on management acceptance. In 1993 USEPA Region 5 staff used the results of a GIS study[31] to examine the relationship of funding and the potential to reduce sediment loading in the Saginaw Bay watershed in Michigan. In 1987 the International Joint Commission for the Great Lakes designated Saginaw Bay as an area of concern (AOC). Designation as an AOC indicates the beneficial uses or biota are adversely affected.

The Saginaw Bay watershed is home to over 1.4 million people and encompasses nearly 8700 mi². It is Michigan's largest watershed and includes all or part of 22 counties. Congressional line appropriations to the state of Michigan provided funding to support planning and implementation efforts within the Saginaw Bay watershed. One major focus of these efforts was the reduction of sediment to Saginaw Bay. In addition to the direct congressional funding support, the USDA had a number of special programs in the AOC watershed to aid in the control of sediment in Saginaw Bay. The goal of the EPA's study was twofold: (1) facilitate agricultural management practice implementation to achieve maximum environmental benefits, and (2) ensure funding was targeted to critical areas needing treatment. The first phase of the project was to estimate sediment delivery and document funding decisions related to sediment control within the AOC watershed. Sediment delivery was estimated for county and subwatershed within the Saginaw Bay watershed using a GIS version of the modified universal soil loss equation (MUSLE). The GIS-generated outputs were field-verified by the Michigan Department of Natural Resources (MDNR) staff. The MDNR–Great Lakes Office used a political advisory committee to allocate its available funds on a political jurisdiction basis, the county. The USDA also used its traditional county delivery system for its sediment control programs. Figure 11.5 shows the fiscal-year 1991 funding and the five counties with highest estimated sediment delivery. The fiscal-year 1992 results were similar to 1991 results. The analysis[32] highlights the ineffectiveness of utilizing the traditional jurisdictional (county) basis to allocate agricultural cost-sharing funds to solve excessive sediment loading problems to Saginaw Bay. Figure 11.5 shows there is no correlation between funding allocations for sediment control and estimated sediment delivery potential. The key conclusion of Phillips' study[32] is, "The sediment delivery and erosion potential for the 22 counties and nine sub watersheds within the Saginaw Bay Watershed indicated you cannot determine funding priorities based on jurisdictions when addressing watershed-scale problems." However, MDNR continued the traditional allocation approach after the study results were released, because of the lack of buy-in by decision makers or the MDNR advisory committee.

Because all files in a GIS are relational, GIS is ideally suited to track land-use and treatment data. The primary advantage is that land treatment information can be spatially displayed and combined with water-quality–related information. A number

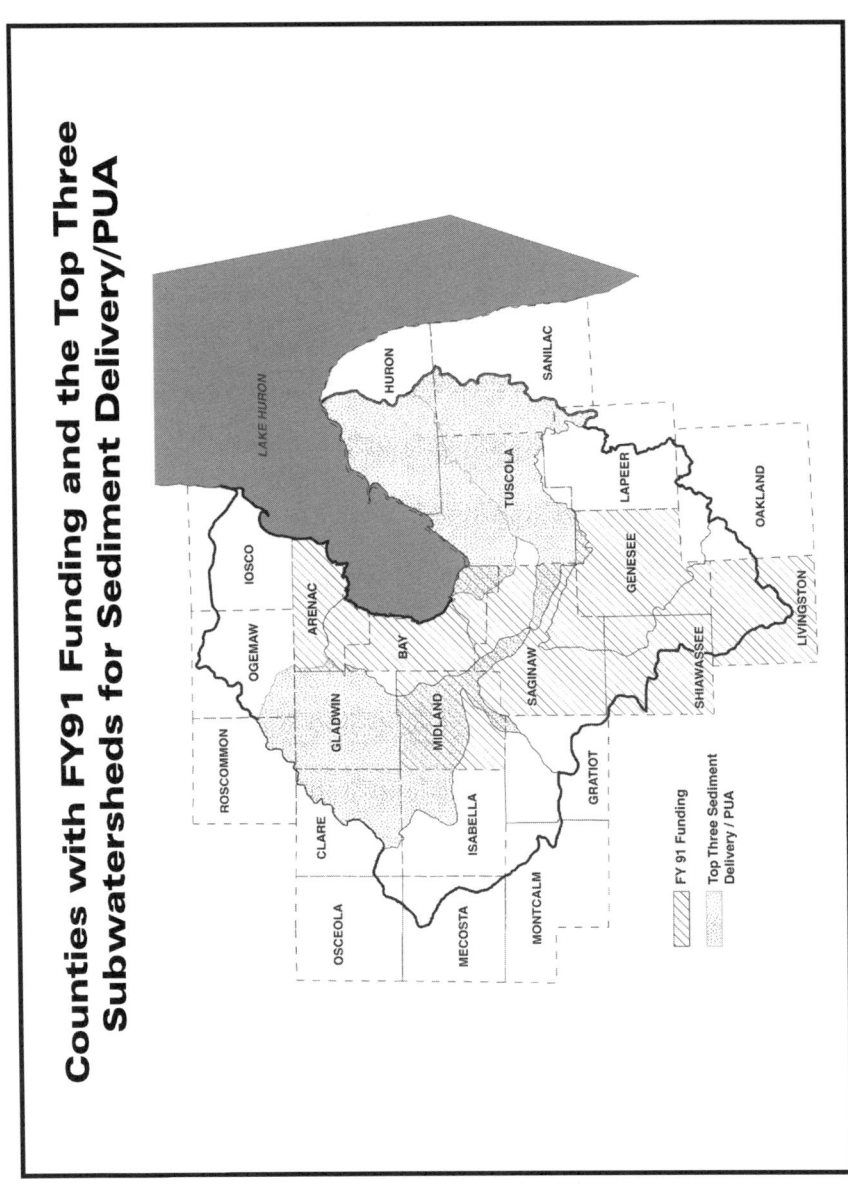

FIGURE 11.5 Counties with fiscal-year 1991 (FY 91) funding and the top three subwatersheds for sediment delivery/PUA.

of rural clean water program projects utilized GIS to display the location of implementation efforts (St. Albans Bay, VT) and highlight critical areas to be addressed through enhanced implementation efforts (Garvin Brooks, MN). The Lake Champlain National Nonpoint Source Monitoring Project utilized a GIS to store and map the watershed's project baseline inventory of soils, topography, land ownership, and agricultural land use; track implementation; and display results.[33] The use of GIS to track land management activity is extremely valuable when evaluating the effectiveness of land treatment and its impact on water quality. This ability to provide a link between management and off-site impacts strengthens the need for close coordination among the implementation teams and the monitoring and evaluation teams.

BOX 11.1
What GIS Is Not

GIS is not a decision-making system. It is a tool that supplies information in a form that allows partnerships to arrive at a decision easier. To utilize GIS-generated outputs, partnerships and technical staff must be aware of the inputs (quality), assumptions made during the data manipulation, and questions the outputs were generated to answer. The maps generated from a GIS are useful to partnerships and technical staff if one ensures that the map meets the intended needs. GIS maps offer watershed communities the opportunity to look beyond town lines to the natural boundary of the watershed. The Massachusetts Department of Environmental Protection (DEP) staff found GIS useful in identifying priority resources to consider when reviewing water withdrawal permits and evaluating potential impacts associated with the withdrawals. The DEP also provided training to communities in the Merrimack River watershed to enable them to integrate a watershed perspective into their planning and regulatory activities.[26] It is important that the benefits of the spatial analyses and mapping do not mask the issues associated with uncertainty, error, accuracy, and precision.

Another major goal of some modeling efforts is to extend water findings over space and time to other projects. Use extreme caution when selecting data sets to utilize in this type of effort. The data must be representative and appropriate. GIS technology was used to estimate phosphorous using a unit-load approach to Owasco Lake (New York); the measured loads for the lake were greater than the estimated loads. Heidtke and Auer[34] determined the difference is due to using sampling results from wet years to develop the measured load values. Any GIS analysis is only as good as the information that goes into it, which is why Phillips[31] field-checked the results to ensure the validity of her conclusions. If a project decides to utilize GIS, a plan should be developed that includes the uses of the GIS, hardware, software, database development and management, data quality standards, staffing, and funding.

For an additional resource on modeling, see USEPA, *Compendium of Tools for Watershed Assessment and TMDL Development*, EPA841-B-007, Office of Water, USEPA, Washington, D.C., 1997.

An additional resource on the use of GIS is Griffin, C.B., *GIS Introduction for Public Agencies*, Conservation Technology Information Center, West Lafayette, IN, 1995a.

REFERENCES

1. TetraTech, Inc., Development of a Simple Nomograph Methodology to Calculate TSS Removal from Urban Runoff, TetraTech, Inc., Fairfax, VA, 1995.
2. Chapra, S.C. and Canale, R.P., *Numerical Methods for Civil Engineers*, 2nd ed., McGraw Hill, New York, 1988.
3. Fleming, G., Deterministic models in hydrology, Irrigation and Drainage Paper #32, Food and Agriculture Organization, Rome, Italy, 1979.
4. Huber, W.C. and Lickinson, R.E., *Storm Water Management Model, Version 4, Part A: User's Manual*, EPA/600/3-88/001a, USEPA, Washington, D.C., 1988.
5. USEPA, *Compendium of Watershed-Scale Models for TMDL Development*, EPA-841-R-92-002, USEPA, Washington, D.C., 1992.
6. USEPA, *Protocol for Developing Nutrient TMDLs*, EPA 841-B-99-007, USEPA, Washington, D.C., 1999.
7. USEPA, *Protocol for Developing Sediment TMDLs*, EPA 841-B-99-004, USEPA, Washington, D.C., 1999.
8. USEPA, *Protocol for Developing Pathogen TMDLs*, EPA 841-B-00-002, USEPA, Washington, D.C., 2001.
9. USEPA, *Basins 2.0 Fact Sheet*, EPA-823-F-99-001, USEPA, Washington, D.C., 1999.
10. USEPA, *National Management Measures to Control Nonpoint Source Pollution from Agriculture*, USEPA, Washington, D.C., 2002.
11. Caraco, D., The Watershed Treatment Model, version 3.0, Center for Watershed Protection, Ellicot City, MD, 2001.
12. USEPA, Coastal Bend Bays and Estuaries Project Progress Report, USEPA Region 6, Dallas, 1996.
13. Richards, R.P., *Estimation of Pollutant Loads in Rivers and Streams: A Guidance Document for NPS Programs*, Water Quality Laboratory, Heidelberg College, Tiffin, 1979.
14. Brown, E. et al., *Developing a Watershed Management Plan for Water Quality: An Introductory Guide*, Michigan Dept. Environmental Quality, Surface Water Program, Nonpoint Source Program Staff, Lansing, 2000.
15. Macatawa Area Coordinating Council (MACC), The Macatawa Watershed Project, Phosphorus Reduction Strategy for the Macatawa Watershed, draft, MACC, Holland, 1998.
16. Mollahan, R., Restoration of the Waukegan River, in *Proc. National Water Quality Project Symp.*, EPA/625/R-97/008, USEPA, Washington, D.C., 1997, pp. 109–116.
17. Cardona, M.E., Stillman, J., and Sinclair, R.A., *Waukegan River Stormwater Quality Simulation: Application of AUTO-QI Modeling*, Illinois State Water Survey, Champaign, 1995.
18. Knisel, W.G., ed., *CREAMS: A Field Scale Model for Chemicals, Runoff, and Erosion from Agricultural Management Systems*, USDA, Conservation Research Report No. 26, 1980.
19. Davenport, T.E., Field Modeling in the Highland Silver Lakes Watershed, interim report, IEPA/WPC/84-026, Illinois EPA, Springfield, 1984.
20. Chandler, R.D., Modeling and Nonpoint Source Pollution Loading Estimates in Surface Water Management, master's thesis, Dept. of Engineering, Univ. of WA, Seattle, 1993.
21. Stallings, C. et al., Accomplishments and challenges of data collection and modeling in North Carolina demonstration watershed project, *Proc. National Watershed Water Quality Project Symp.*, EPA/625/R-97/008, USEPA, ORD, Washington, D.C., 1997, pp. 243–247.

22. U.S. Dept. of Interior–Geological Survey (USDI-GS), Evaluating the Effects of Urbanization and Land-Use Planning Using Ground-Water and Surface-Water Models, USGS Fact Sheet FS-102–01, USGS, Middleton, 2001.
23. Conservation Technology Information Center (CTIC), 2000 Crop Residue Management Survey, CTIC, West Lafayette, IN, 2000.
24. Gordon, S.I., *Computer Models in Environmental Planning*, Van Nostrand Reinhold Company, New York, 1985.
25. Walsh, S.J., Geographic information systems for natural resource management, *J. Soil and Water Conserv.*, 202–205, 1985.
26. DeShazo, R.P. and Garrigan, P., Merrimack River Initiative Watershed Connections, Lessons Learned in Subwatersheds of the Merrimack River Watershed: The Nashua, Souhegan and Stony Brook Watersheds, New England Interstate Water Pollution Control Commission, Wilmington, 1996.
27. Griffin, C.B., *GIS Introduction for Public Agencies*, CTIC, West Lafayette, IN, 1995a.
28. USACERL, *GRASS 4.1 User's Reference Manual*, U.S. Army Corps of Engineers Construction Eng. Research Lab., Champaign, IL, 1993.
29. Griffin, C.B., Data quality issues affecting GIS use for environmental problem-solving, *National Conf. on Environ. Problem-Solving with Geographic Information Systems*, Cincinnati, OH, Sept. 21–23, 1994, pp. 15–30, USEPA, Washington, D.C., 1995b.
30. Chen, C.W., Herr, J., and Weintraub, L., Lessons learned from stakeholder process, in *TMDL Science Issues Conf. Proc.*, St. Louis, Mar. 4–7, 2001, pp 77–85, Water Environment Federation, Washington, D.C., 2001.
31. Atlantic Research Corporation (ARC) and USEPA, *Saginaw Bay Watershed Sediment Delivery and Erosion Potential GIS Study*, USEPA, Chicago, 1993.
32. Phillips, N.J., Saginaw Bay Sediment Funding, staff report, USEPA, Chicago, 1993.
33. Meals, D.W. and Ferree, C., Project Baseline Inventory: Lake Champlain Basin Agricultural Watersheds Best Management Practices Implementation and Effectiveness Monitoring Project, School of Natural Resources, Univ of VT, Burlington, 1995.
34. Heidtke, T.E. and Auer, M.T., Application of GIS-based nonpoint source nutrient loading model for assessment of land development scenarios and water quality in Owasco Lake, New York, *Water Science & Technol.*, 28(3–5), 595–604, 1993.

Table 11.1. USEPA, Report of the Federal Advisory Committee on the TMDL Program, EPA 100-R-98-006, USEPA, Washington, D.C., 1998.

12 Social Capacity Building

Everything has changed, but our way of thinking.

— Albert Einstein

12.1 INTRODUCTION

The focus of this chapter is on one aspect of the overall watershed management effort: building social capacity. Building social capacity is providing the foundation to support behavioral change within a watershed. Watershed management is about working with people to make the right short- and long-term decisions in their daily lives. In most watersheds, this means changing residents' perceptions and behavior. Many types of changes precede behavioral change: perception of the problem, awareness of the resident's relationship to the problem, knowledge of what residents can do, and creating the desire to change. Successful social capacity building efforts move stakeholders through various stages of awareness, knowledge, understanding, ability, and desire to active participation. It empowers citizens and helps them understand that they can have positive effects and influence in their community. In addition to the change process, this component provides support to ensure the behavioral change is sustained.

The specific watershed management plan component that contains the social capacity building activities is called outreach. Watershed management efforts must motivate stakeholders to accomplish short- and long-tem goals. Watershed management plans' outreach efforts include information and education as well as public involvement and communication. While some of these terms have historically been used interchangeably, for the purpose of the recommended watershed management process they are viewed as quite different. The key difference between information, education, and outreach is the degree of involvement of the participant or the stakeholders. For the purpose of the "four-step process" the following distinctions are made:

The concept of *information* is twofold: public information and support information. Public information involves the distribution of information on specific issues either directly or indirectly to the public. Public information efforts are passive and can be frustrating without the necessary support system. The stakeholders learn about the problem but receive (1) no specific direction about what they need to do to solve the problem, and (2) usually insufficient information to make up their mind about solutions. The information may or may not suggest general actions for the stakeholders. The focus is on awareness and reporting facts, providing the basis for individuals to get involved and make decisions. These types of efforts rely on fact sheets, public service announcements, informational meetings, posters, newspapers, television and radio, exhibits, newsletters, presentations, direct mail, signs, and brochures. Support information involves providing specific technical and management information to targeted audiences in direct support of education efforts or as a follow-up to implementation. This helps the participant operate and maintain the management practice he or she has implemented.

The concept of an *education* program is a hands-on process. This type of effort directs the participant toward developing active problem-solving skills that enable the participant to make an informed decision and take action. The education efforts rely on participants obtaining basic knowledge and active involvement such as hands-on training and demonstrations. This includes one-on-one sessions with landowners and operators. An example of institutional educational is the Maine Nonpoint Source Training and Resource Center. It delivers technical information to professionals involved in activities that may contribute significant pollution. It provides a mechanism for agencies to provide or coordinate technical training to professionals involved in potentially polluting land-use activities. The Center hosts and cosponsors training and provides videos and a lending library.

The concept of the *outreach* program integrates public information and education programs and is a hybrid of the two. Outreach results in problem-solving actions. Madame Chiang Kai-Shek said, "I am convinced that we must train not only the head, but the heart and hands as well." This hybrid relies on information to build awareness of the issue and the need to change behavior or implement management practices and education to provide the opportunities and skills to make the change and thus make a difference. The outreach hybrid is best summarized as, "Tell me, I forget; show me, I remember; involve me, I understand" (an ancient proverb passed to me by Walt Poole). Successful outreach programs get people excited by making it interactive, informative, and fun.

In this watershed management process *public involvement* (participation) is defined as the activities that enable the public and stakeholders to become aware and participate in the process. It includes information, education, and outreach activities that are either general or targeted to specific audiences such as cattle ranchers. When it comes to watershed management, stakeholders need to understand that the effects of their activities along with the decisions and actions of many people contribute to the cumulative impact of pollution and natural resource degradation in a watershed. The various components are utilized to make sure everyone involved knows the basics. Public communication is used to inform the public of issues, events, and opportunities to participate and report progress. It is specifically related to the watershed management process, and its focus changes with each phase and includes a pubic information component.

Through the use of information, education, and outreach programs, people are provided an opportunity to increase their understanding of the watershed they live in and how their actions affect its water quality and natural resources. High-quality information, education, and outreach efforts can facilitate changes in the watershed residents' lives that directly impact the quality of their water resources. Many adults did not learn much about watersheds in school, and information about their specific watershed may not be widely available. If a partnership fails to provide opportunities and information to the public, then the partnership fails to provide the means for people to change small daily activities that together result in poor water quality and natural resource condition. In order to build public support and make a difference, partnerships need to explore ways of "making watershed management plans and assessments available to the public." Public information, education, and participation programs give residents the information and knowledge to be proactive and responsible. These kinds of programs can effectively build public support for regulatory programs, when and if regulatory measures are necessary to protect and restore water quality.

12.1.1 PUBLIC AND STAKEHOLDERS

The word "public" implies a uniformity that is not there. For this reason, public involvement needs to be thought of in terms of "publics" – groups of individuals who share the same type or degree of interest in the issue. At a minimum "publics" include:

Individuals or groups known to be affected by the issue
Individuals or groups who might be or who think they might be affected by the solutions
Individuals or groups whose support is necessary to enact the solutions
Individuals living in the watershed who are unaware of a direct link to their watershed

The three groups clearly will differ in many of their perceptions, attitudes, and behaviors with respect to a watershed, and these variances need to accounted for in facilitating their participation in the watershed management effort. In chapter 4, on partnerships, the first three categories were utilized to define stakeholders' categories. Stakeholders have a knowledge or direct interest in the watershed. Another way to categorize the nonstakeholder public is in four groups: politicians, decision makers, activists, and ordinary citizens.

12.2 OUTREACH

The watershed outreach strategy consists of an education component, communications plan, and public involvement component. It should cover the five Ws: who, what, why, where, and when. Riley[1] identifies common features among several outreach strategies found in different communities for developing support for restoration projects; schools or universities are involved, volunteer tree planting or cleanup projects are held, workshops or conferences are held, success celebrations are held, politicians are invited and involved, and press coverage is sought. A well-developed outreach strategy also cultivates the kind of environment more accepting of change and innovation.

The outreach strategy provides the foundation for changing the people's behavior in a watershed. The strength of this effort will depend on how well the strategy is linked to the goals of the watershed management plan through measurable objectives related to behavior change and the needs of watershed residents. The watershed management outreach component is not one isolated set of activities; it is a number of small projects aimed at different audiences for different purposes to achieve an overall goal. These activities will be based on the goals and objectives for each target audience in the watershed. The outreach strategy will help take individuals through the stages of knowledge, persuasion, decision, implementation, and confirmation.

After establishing an outreach team or committee, the hard part begins — to set an acceptable planning and design process. Press coverage can be effective in reaching watershed residents.[2,3] The media should be invited to join the partnership's outreach team (depending on the structure it could be a committee, subcommittee, or workgroup). The following four steps will help guide the outreach team in the development of an outreach strategy that has a purpose for all watershed residents.

1. Determine target audiences. The outreach team along with the planning committee should make a list of the watershed communities and potential target audiences. Understanding the communities in the watershed is necessary in order to evaluate existing resource conditions, causes and impacts, and assess the effects of alternatives, including effects expected if the watershed remains untreated. Understanding communities is critical during the planning and implementation processes. Socioeconomics affect the planning and implementation processes and their outcomes, generally influencing conflicts, cohesion, and public involvement during the planning phase; decisions about management alternatives; and whether and how the watershed management plan is implemented. Frequently target audiences are divided into individuals who need to change for environmental improvement to occur. Individuals who can support change directly through affiliation or indirectly such as teachers are one such audience. During this step a needs assessment should be conducted. Needs assessment covers existing knowledge of problems and issues, perceptions about water quality and other issues, barriers that prevent the messages from reaching the target audiences, how they access information, and who or what they consider a reliable source of information. The overall needs assessment will provide the information necessary for targeting priority audiences with outreach activities and specifying how messages need to be structured to be effective.
2. Determine outreach goals. The outreach goals should focus on target audiences' being aware of the water resources and their uses, knowledgeable about issues and problems, and being able and willing to implement solutions. Goal setting is important as it is necessary that the target audiences attain a specific level of awareness, knowledge, and skills in order for a watershed management project to meet its water-quallity goals.
3. Develop objectives. While the goals prescribe the future condition, the objectives give the outreach strategy focus for each target audience. An objective is a statement of what a target audience should know, skills they

should acquire, or practices and actions they should take after the outreach activities. The outreach team needs to identify what individuals need to know and be able to provide the necessary information and education support for them to make a decision. The team needs to think in terms of who — farmers, municipal officials, joe public, fishermen — will be able to do what — take action. Objectives need to be specific enough to be measured to determine if they are being met, and if not, why. Evaluating the achievement of measurable objectives is an integral component of the watershed management project's overall evaluation.

4. Determine appropriate activities. Develop a list of activities and actions that must be done to achieve the determined objectives. Prioritize activities and determine who is responsible for ensuring completion (it is extremely important to identify the cooperators and partners for each activity; evaluate criteria and include a schedule and budget). The actual implementation plan is the road map for the effort. It provides the what, where, who, why, how, and when of the effort. Table 12.1 shows a framework (action plan) that can be utilized for this step.

The action plan (Table 12.1) translates the implementation plan into definable work items. The action plan provides the basis for a day-to-day schedule, details of the outreach strategy, including such things as the tasks that need to be accomplished, the materials that should be developed and provided, the events that should be attended, the organizations with which to partner, and the timing for communication.

The Three Rivers Project (Ireland) implemented a public awareness campaign with the primary aim to inform all sectors of the community about their responsibility in preventing water pollution. From the beginning, the outreach effort addressed two challenges: first, adults learn differently from children; and second, rural landowners, the primary audience, tend to be resistant to change. Swain highlights that rural landowners are more resistant to change and need more justification and capacity building to make changes.[4] The public awareness campaign is a very important part of the Three Rivers Project and is managed at two levels: an overall media campaign developed for the general public and targeted campaigns developed for farmers and school-aged children. The general campaign's goal was to exploit major events within the project period to gain maximum media coverage to promote the project objectives. The initial focus of the public awareness campaign was on the farming community, to encourage participation in project pilot areas. Farmers were informed about the project through a brochure, a series of public meetings, and a number of press releases. The approach proved successful as documented by a high percentage of farmers in the pilot areas becoming involved in the project.[5] The other targeted campaign was for primary schoolchildren. It was thought that one of the most important aspects of the public outreach campaign was to involve the children of the families in the watershed. The Three Rivers Project developed a unique "Happy Fish" campaign aimed at educating schoolchildren and their families about the local water-quality problems. In addition, the overall awareness effort included demonstrations, web pages, and meetings with other community groups.

TABLE 12.1
Example of an Outreach Action Plan

Objective	Activity	Lead	Partners/Cooperator	Schedule	Cost	Evaluation Criteria
In 2002, increase awareness of nutrient management in the Emily Creek watershed by 10%	Farmer workshops on the economic and environmental benefits of nutrient management	NRCS	Extension service, co-ops, crop consultants	Spring	$1500	Percentage of farmers from the Emily Creek watershed who participate in the workshops
Increase the number of urban residents that implement streamside buffers	Individual land owner install streamside buffers.	City Public Works Department	Fish and Wildlife Service, Department of Natural Resources	June 2002 to June 2004	$80,000 for technical assistance; $15,000 outreach; $240,000 incentives	Number of parcels with streamside buffers

Social Capacity Building

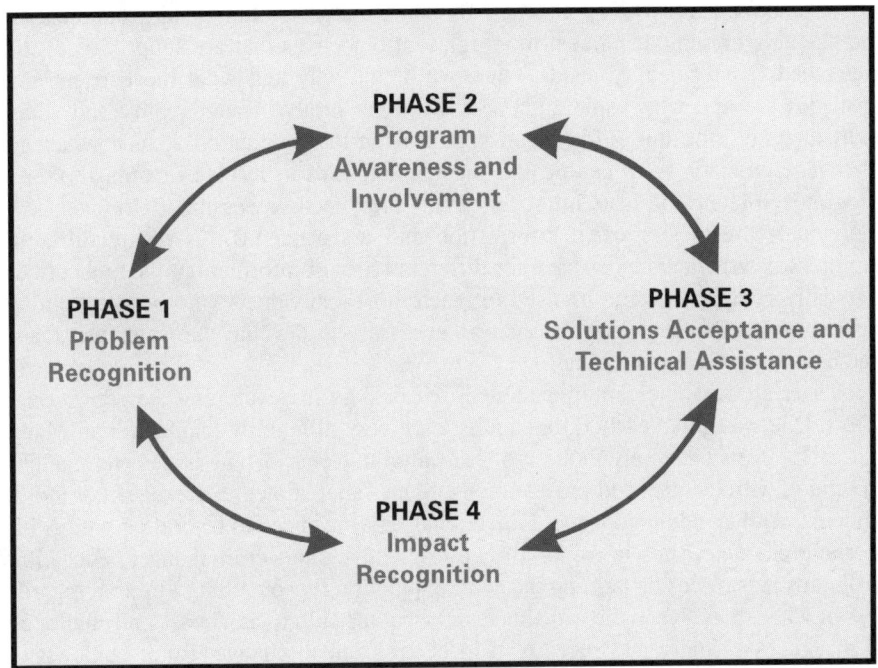

FIGURE 12.1 The outreach strategy and its relationship to the four-step process.

12.3 EDUCATION

The educational activities of the outreach strategy are an integral part of a watershed management effort's success. Figure 12.1 shows the four-phase process or cycle and its relationship between the phases and the outreach strategy. Educational efforts are cumulative and become the glue that holds the partnership and process together. For example, stakeholders must be educated to first realize the actual and threatened problems exist and then to work with the planning committee to develop and implement solutions. Stakeholders are also crucial for working with funding organizations to allocate financial assistance and staff to solve the problem. Without the right outreach effort, the steering committee cannot adopt workable solutions to the problems or ensure stakeholder buyin is sufficient to guarantee that solutions are implemented.

The severity of water-quality problems in a watershed is considered a symptom of the combined actions or lack of action of all the stakeholders who use or own land in the watershed. The challenge facing watershed partnerships is to look beyond assumptions about what people are or are not doing, but at what they can do. As mentioned earlier, "needs assessment" is one tool that can be used to find out exactly what the land uses/management activities are and why individuals do what they do. A "needs assessment" is important because it can make a watershed management plan more effective and efficient by focusing on overcoming the barriers and obstacles. A better understanding of the target audiences helps define the delivery mechanisms and the content of approaches. By better understanding the target audiences, the planning committee can set more relevant objectives that are supportive of the watershed's water-quality needs. Once we define why and what the barriers and obstacles are and why some audiences willingly protect water quality and some individuals do not, this information will provide the foundation against which to plan. The outreach team can design specific activities to resolve or remove them. Program efficiency is also improved when the "needs assessment" defines how individuals prefer to receive information and assistance. By assessing different preferences target audiences have for different types of information, the most effective delivery approach can be used to reach those individuals who need the information most. This provides information necessary to develop the watershed management efforts' campaign plan.

There are two interesting challenges to consider in developing the educational effort. One was noted earlier, that adults learn very differently than children. Many outreach efforts have only thought of education in terms of how issues were taught in school, with lectures and presentations. Many other strategies are as, if not more, effective with an adult audience. Educational activities need to be crafted with adult learning characteristics in mind. To cause effective short-term changes, education programs must focus on helping the adult population try something different regardless of where they live in the watershed and what the culture is. The second challenge is that many primary audiences tend to be resistant to change. Rural landowners need more proof of the need for change and tend to hold on to old habitats and beliefs longer than many urban residents.[4]

The most effective education approach with adults is one-on-one. The RCWP found that one-to-one contact appeared to be a necessary condition for project success. It was not sufficient if landowners did not support the goals and objectives of the project.[6] In a survey, Simpson reported that adults in Milwaukee were willing to learn and make personal lifestyle changes to help the water environment, but they preferred a passive approach (where everything is delivered to them) to the education process.[3] When providing one-to-one contact with stakeholders, the role of the implementation team will vary by situation and stage in the idea adoption process. The following is a list of common roles implementation team members will have to fill and associated behaviors:

Educator: provide skills and knowledge to stakeholders
Advisor: provide data and advice
Enabler: help stakeholders make a change in behavior, adopt a management practices, etc.
Encourager: actively support individuals who are adopting the necessary measures
Advocate: committed to the cause and work to mobilize other stakeholders

Individuals adopt new ideas in four stages. The watershed educational effort must be flexible enough to provide the proper role and appropriate support to ensure the adoption of the desired behavior. Culture shapes what a group of stakeholders defines as relevant knowledge concerning how they manage their land, and this will directly

Social Capacity Building

FIGURE 12.2 Adopting new best management practices. (Courtesy of CTIC, West Lafayette, IN.)

affect which knowledge an implementation team must focus on. Knowledge (ideas) from the external environment is expected to be a starting point, not the end of behavioral change. Figure 12.2 shows the dynamics of the adoption process where the stakeholder combines inputs from the implementation team to make a decision regarding modifying behavior. The external inputs must be provided in a manner that acknowledges the stakeholders' culture and the existing site-specific management conditions. The partnership through the outreach and implementation teams must inform stakeholders before they ask for change. Stakeholders have to do something different, change behavior, implement new environmental practices and behaviors — some of these may be inconvenient — rather than continue present behavior and habits. The adoption process must support overcoming this inconvenience.

12.3.1 KNOWLEDGE STAGE

What is the idea/concept? How does the concept work? How well do I understand it? How will the stakeholder gain an awareness of the issues and ideas? Figure 12.3 highlights the cultural elements that influence behaviors.

> Education effort: the idea/concept must be clearly communicated so the individual can understand how it works and what it means to him or her. Public meetings, forums, and workshops on key issues are held. Public field trips to build awareness by showing problems and solutions are held. The edu-

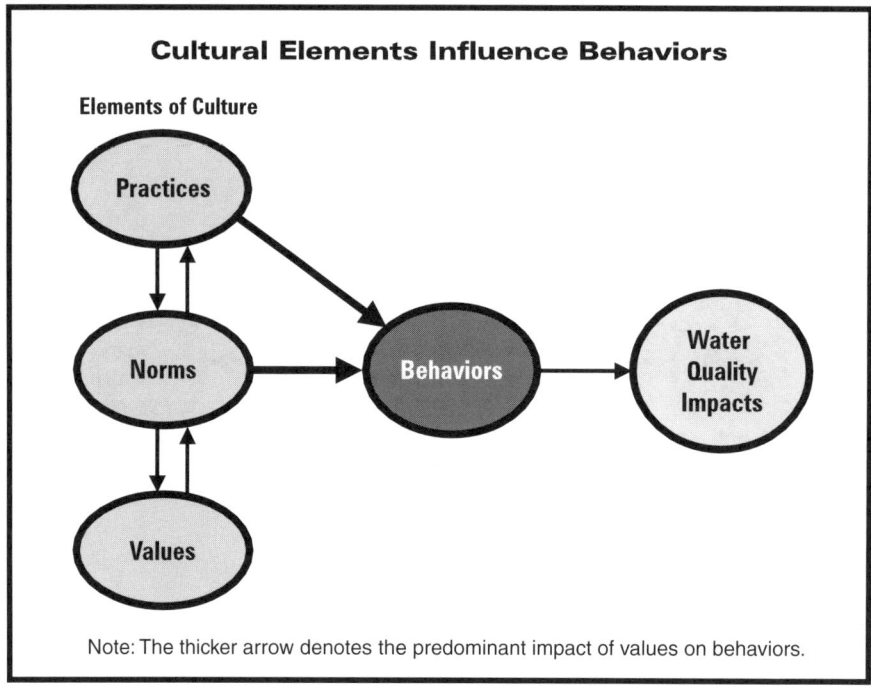

FIGURE 12.3 Cultural elements influence behavior. (Courtesy of CTIC, West Lafayette, IN.)

cation effort should be closely linked with the information effort to provide support for creating this awareness.

Implementation team role: educator, advisor.

12.3.2 Persuasion Stage

Why do it? What is the advantage of the idea? What are others doing with the concept? Should I do it? The individual stakeholder understands the idea at a gross scale and is developing an opinion about the idea.

Education effort: provide information on available technical and financial assistance. Ask participants to talk with their neighbors about the project and why they decided to become involved. Provide names of participating landowners already utilizing the approach being promoted, so neighbors can teach neighbors. Ensure the right information is provided and work with the target audience on its interpretation. Information to individuals must be delivered through the appropriate mechanisms and with enough detail to support adoption of the desired behavior. If the target audience is moved from awareness of the problem to personal readiness to take specific actions, the education program is a success.

Implementation team role: advisor, encourager.

Social Capacity Building

12.3.3 IMPLEMENTATION STAGE

I'll do it, now what?

> Education strategy: provide necessary information and opportunities so the individual can follow through. Workshops on key installation and operation and maintenance issues are held. Ask participants to talk with their neighbors about the project and why they decided to become involved. Hold field trip and demonstration days to show stakeholders what success will look like.
> Implementation team role: advisor, enabler.

12.3.4 EVALUATION (CONFIRMATION) STAGE

Did it work? Did it have the desired impact? Did I make a difference? Should I continue to do this, something new, or go back to the original activities?

> Education strategy: do follow-up with and help individuals see what their actions mean to the overall watershed.
> Implementation team role: encourager, enabler.

Rawls Creek in South Carolina was recently the focus of a watershed education and assessment project designed to identify the nonpoint sources of bacterial pollution and increase public awareness of the runoff pollution and personal behaviors that needed to be changed to reduce the pollution. Results from the assessment study suggested that the impervious areas were the greatest source of fecal coliform bacteria in the watershed. The Rawls Creek project then targeted youth organizations, landowners, and community leaders to raise awareness of the polluted runoff problems and related land-use issues. The 2-month program directly involved almost 500 individuals in seminars, community meetings, and service projects. Everyone who participated in the educational program was informed about watershed, land-use impacts on streams, stormwater, and polluted runoff. All participants were encouraged about their newly enhanced awareness to the community service project. Seventeen groups, including families, school classes, and scout troops, tagged approximately 700 storm drains and provided over 3000 residents with information about stream protection and polluted runoff.[7] Rawls Creek shows the multiplier effect education can have in a watershed project.

12.4 COMMUNICATIONS PLAN

A proper communications plan provides more than another tool to address watershed management issues; it helps create the right environment. Effective communication is clear, concise, and consistent. The communications plan must be structured around these three concepts. The traditional questions of who, when, where, and how are a reasonably good starting point for developing a communications plan. With whom will the partnership communicate? Consequently, a strategic question is: Who is empowered to talk about what? Communicators explicitly and implicitly choose what to talk about, and what to ignore. How the agenda is shaped can profoundly impact the partnership's communications with stakeholders. The communications plan needs to promote external communication and define the internal communication process. Communications plans not linked to underlying watershed management plan goals are as effective as an aesthetically pleasing advertising campaign that fails to generate business. The communications plan provides the framework for the distribution of the message.

A communications plan provides the basis for structuring, executing, and evaluating communication practices. Communication is a two-way process. All the traditional communication forums — newsletters, meetings — naturally flow from the plan. Technologies like the Internet should also encourage an outreach team to reconsider traditional top-down communication strategies. Clarity of purpose can transform the communication network while significantly improving partnership performance. The communications plan can be developed either deliberately or by happenstance. Thoughtful analysis of the partnership's communication needs and stakeholders' concerns can help the outreach team make the appropriate choices and tradeoffs that result in an effective plan designed to incite meaningful actions. There is no one-size-fits-all communications plan, because partnerships have different objectives. A successful communications plan is as much about what is not said as it is about what is said. The communications plan has a great deal to do with how events will be remembered, which in turn shapes the public's responses.

Developing the communications plan is a basic step in establishing a well-designed and effective strategy to communicate with the "public." The formation of a comprehensive communications plan provides a foundation that will ensure the watershed plan's eventual implementation will be successful. The plan needs to answer:

>How will the partnership and stakeholders communicate?
>When and why will stakeholders and the partnership communicate?
>In forums will stakeholders and the partnership communicate?
>Partnerships need to ask a fundamental question: when is speed more important than comprehensiveness?

We must address many tension points, but a fundamental strategic issue question persists: how do partnerships make the appropriate tradeoffs between too much information and too little? Another critical issue is how to properly balance why and what messages should be sent from the partnership to stakeholders.

A number of difficulties pervade partnerships seeking to determine the objectives of a communications plan. Many do not think about their communication objectives explicitly. Others have a vague objective of keeping everyone informed, but no idea about what, in how much detail, in what way, or how often.

The Communications Plan
The messages from the partnership should be defined in the context of the watershed management effort. In general, the messages should use well-known, nontechnical terms. The plan should use short and fewer messages to gain attention and interest.

The communications plan can be viewed as a broadly defined strategy at the macro-level where the steering committee makes choices and tradeoffs between proposed activities, based on the organizational goals and judgments about expected reactions from the public; it then serves as a basis for action. As with all planning efforts, the goals and objectives of a communications plan provide the framework for all the steps that follow. The communications plan's overall goal needs to be fairly broad, yet should encompass the very essence of the partnership and its relationship to the development and implementation of the watershed management plan. The goal should reflect the desired outcome of the communication effort. For example, "To create in watershed residents awareness of the Lake Emily Watershed and the water cycle, which will lead to a change in watershed residents' habits that reflects a concern for water quality." The goal can be viewed as the mission for the communications plan. The objectives must be specific enough to narrow the focus of the activities; these objectives must be measurable and have a single desired outcome. For example, "increase urban residents' awareness and knowledge of the watershed and subwatershed in which they live." The objectives need to be prioritized to enable the action plan to be developed and implemented in a systematic manner.

Typically an effective communications plan will be a mixture of various strategies. The strategy for a particular issue depends on the target audience, message, and expected response. The following is a list of typical communication strategies:

Spray and pray. This strategy is based on the idea that the steering committee should shower stakeholders with all kinds of information.

Tell and sell. This strategy has the steering committee communicating a more limited set of messages and information that they believe address core partnership issues. First, tell stakeholders about the key issues. Second, sell the stakeholders on the wisdom of their approach.

Underscore and explore. The steering committee focuses on several fundamental issues clearly linked to the partnership success, while allowing the outreach team the creative freedom to explore the implications of those issues in a disciplined way.

Identify and reply. This approach focuses on stakeholder concerns. It stresses the importance of making sense out of the often-confusing implementation process and status.

An essential step in the development of a communications plan is understanding the "publics" that will be the focus of the effort. The various characteristics of the identified publics need to be documented, especially the socioeconomic traits. These traits will provide an understanding of the marketplace and background for determining the need for and characteristics of financial incentives. By understanding the marketplace and its potential impact on the development and implementation of the proposed watershed management plan, the communications plan can be utilized to target efforts to specific audiences during windows of influence. *Managing Lakes and Reservoirs* compares the strengths and weakness of four social survey techniques: mailed questionnaire, telephone interview, in-person interview, and focus groups interview.[8] The analysis of the four techniques shows a relative range in cost, timeliness, and response rates, indicating a partnership could match its needs and constraints to a particular technique.

The Huron River Watershed Council[9] outlined five steps to follow in implementing market research to support watershed management:

1. Identify the subjects from whom you wish to get information.
2. Identify the communities you want to target.
3. Design research surveys to gain the desired information.
4. Conduct the surveys.
5. Compile and analyze the data.

While broader than a "needs assessment," a market research survey can provide the basic information needed. Before conducting a survey, the outreach team needs to decide what will be done with the information. For example, will it help identify basic environmental educational needs in the watershed? Determine a baseline on stakeholder attitudes? Answers to these types of questions will help formulate the survey. When the team develops the survey, test it on a small group of stakeholders to make sure the survey understandable. When conducting the survey, the surveyor must clearly identify the partnership, the purpose of the survey, and who he or she is. The surveyor should explain how the partnership will use the information and ask the participants if they want a summary copy of the completed survey.

The Huron River Watershed Council learned an important lesson for conducting surveys. The Council[9] attempted to use volunteers to complete a survey. The lack of organization and failure to provide the necessary support to ensure the volunteers were successful led to a small fiasco in conducting the survey. Months later, learning from the previous experience, the Council was better organized to support the effort. One major change was the identification of discrete tasks they wanted the volunteers to work on. This resulted in much greater success in conducting the survey, and in addition it strengthened the role of volunteers in the overall effort.

Market research will provide definition to the target audiences for the communications plan. A target audience is a group of watershed residents or stakeholders who have common characteristics. The most common ways to define a target audience are demographics, behaviors, and attitudes. Demographic criteria mostly commonly used are occupation, household types, income, geographic location, and education. Behavior or activities that have a negative or positive impact on the watershed, types of media they respond to, organizational membership, attitudes or perspectives on watershed issues, beliefs about an individual's impact on the water quality, hobbies, and the role of government in watershed management are also used to characterize individuals. Define the target audiences in the watershed in terms unique enough so the implementation team can target them in the communications plan, but broad enough so every resident is not considered an individual target audience. If done correctly, specific target audiences will correlate highly with particular behaviors. The outreach team needs to decide if it will target audience or behavior. Targeting behavior focuses messages on impacts of behavior regardless of socioeconomic characteristics of the individual. Instead of focusing on teaching people that they cause problems, which engenders guilt, communicate that the

Social Capacity Building

individuals are capable of improving the waterbody, which is empowering. The link between the individual's behavior and the environmental impact is weaker for a target audience approach since the individual might not recognize he or she exhibits the behavior in question. But knowing from whom (target audience) the implementation team is expecting participation or behavior change allows the outreach team to focus more accurately on the desired behavior.

The outreach team must define the following aspects of behavior change:

- How does the target audience's current behavior adversely impact water quality?
- How does this current behavior relate to the goals of the plans?
- What desired behavior needs to replace the current behavior?
- What is the water quality and social impact of the desired behavior?
- Is the desired behavior enough or the start of a long-term process of social change?
- What incentives does the target audience need to switch to or implement the desired behavior?

Since the partnership deals with "publics," it is important to prioritize the target audiences/behaviors that will have the most impact on meeting the watershed management plan's goals. The needs assessment can determine the necessary behavioral change, which is factored into the objectives set for the communications plan and directly related to the goals and objectives of the watershed management effort. The market research defines the message to reach them. When evaluating the target audience for prioritization, the outreach team must assign a level of effort to be directed toward each target audience based on the relative importance to the watershed management plan's goals and the intensity of the behavioral change needed. The prioritization will change over the life of the effort, as objectives or phases or both are completed or achieved, and the level of effort needs to be adjusted based on the resources available at any time.

Overall watershed management process messages should answer the following questions:

- What is the problem or issue?
- How does it affect the target audience?
- Why should the target audience care?
- What can the target audience do?

The stakeholders are the partnership's lifeline. Communicate with them and through them, and build on the strength of these relationships.

For example, a watershed effort focused on addressing the issue of excessive wastewater impacts would have two audiences. The first audience would be watershed residents, and the message would focus on the overall impact of wastewater impacts and their sources. The second target audience would be septic system owners; the focus would be on what a poorly managed or sited septic system contributes and what the audience can do about it. A community with a septic system with a suspected problem or a concern would be encouraged to contact the local health department of the watershed partnership for assistance. The landowners whose wastewater is handled by publicly owned wastewater treatment plant would be excluded from the septic system outreach efforts. Focus on keeping messages simple and straightforward, and communicate only one or two messages at a time. Use graphics and pictures to illustrate the message where possible, because a picture is worth a thousand words.

12.5 PUBLIC PARTICIPATION

Public participation is any process that involves the public in problem solving, planning, policy setting, or decision making. It uses public input to make decisions and allows stakeholders to influence and share responsibility for decisions. Public participation means an exchange of concerns, information, ideas, and preferences related to the issue under consideration. Increasingly, watershed partnerships and project sponsors are recognizing that the appropriate "publics" must be involved in the decision-making process. The benefits of public participation are well known and include the following:

1. The decision takes into account the publics' perspectives, values, and knowledge of the issue and possible solutions.
2. Usually a stronger commitment results when the public has been involved in the decision-making process.
3. Involving the public in decision making invokes volunteers to become involved in the implementation of the watershed management plan.

Successful public participation in the decision-making process for watershed management planning and implementation phases needs to be based on the following principles:

1. There are many publics; none must be excluded.
2. The earlier, the better for initiating public involvement.
3. Public involvement must continue throughout the entire watershed management process.
4. Clear, accurate, and timely information about the watershed management process and progress must be provided to the public throughout.
5. The public participation process must be clearly laid out concerning scope, intent, and schedule.
6. Public input must be treated professionally and with courtesy, and public input must receive appropriate feedback.
7. Public involvement must be continuously tracked and evaluated to ensure the effort is current with the process.

Due to its importance, a separate section of the watershed plan is recommended to describe the activities that will be used in the watershed management effort to obtain and maintain local support for the effort. This is especially important for watershed management efforts based on implementing management practices on individual property owners' land. Involvement of the public in the watershed management process is necessary for the planning committee to make better decisions regarding the management approaches. Public involvement results in decisions that have the support and commitment of the community, are responsive to local needs, and reflect the desires of the community. Public participation also contributes the direct, immediate knowledge of community members about the watershed conditions, concerns, and issues. It reduces the likelihood of conflict, legal action or delays, and expenses by incorporating local issues and concerns into the assessment and planning phases. Public participation increases the potential for plan implementation by demonstrating broad community support for its action items.

Local coalitions and participatory processes are vital to motivate local governments and citizens. Emphasize influential and voluntary participation of multiple local and nongovernmental interests. New watershed initiatives involve partnerships between public and private interests. Levels of participation are likely to vary over time, with ebbs and flows in intensity. So an ongoing outreach effort is needed to ensure effective participation, improve citizen awareness of issues, and increase public understanding of what needs to be done. It is easier to build public consensus for action when informed people believe they are protecting or restoring a particular water resource, especially one that is near and dear to their hearts. Local educational activities should be planned with participants and partners whose mutual intent is achieving outcomes that have impacts. Get schools involved, and explore the possibility of linking the watershed management effort with the GLOBE program for schools. Kids will jump right in anyway, and involving them creates an incentive to do it right and provides a basis for watershed education in the school system. The Darby-Cobbs watershed partnership's public participation committee formed an education subcommittee to work with learning programs in municipal school districts and identify existing educational resources and suggest the creation of resources that did not exist. Through a brainstorming process ten projects were identified for implementation:

1. Produce a watershed status report based on initial technical committee reports.
2. Conduct a resident survey of issues and awareness.
3. Hold an educational symposium.
4. Develop a watershed awareness video and public service announcements.
5. Develop other educational/promotional materials.
6. Provide targeted materials and workshops for the municipal audience — perhaps something about Phase II storm water regulations, management practices, storm water ordinances, and so on.
7. Develop a watershed web site.
8. Collect or create educational materials for municipal officials regarding watershed management and tools to improve watershed stewardship.
9. Create or provide access to teacher training opportunities.
10. Facilitate service learning.[11]

The watershed management plan provides the basis for marketing the effort. The partnership believes it will increase membership, help raise funds and other resources, attract the media, and build credibility.

Center some partnership outreach efforts around events. A watershed fair to educate the public on development of the watershed management effort and how it affects them, solicit public input, increase public awareness of watershed issues, and explain links to land use is an ideal watershed event. A well-planned watershed conference with public participation can be an important step in gathering public support for a watershed management effort.[2] Figure 12.4 is an example of a newspaper add soliciting participation in a project. Another excellent opportunity for showcasing the watershed project comes with media events that occur annually such as national celebration for Wetlands Month, National Drinking Water Week, and Lakes' Appreciation Month. The important thing is to get lots of press coverage for the partnership's efforts. The outreach team can organize tours and host receptions to get the word out.

Outreach in support of public participation is critical at the beginning of the watershed planning effort, during the planning process and implementation phase, and after implementation is completed. Typical "publics" for watershed management projects are local, state, and federal government agencies; environmental and conservation organization; individuals living or working in the watershed; businesses in the watershed or that rely on material from the watershed; taxpayers; and national environmental organizations. Although public participation is important, historically the local community has often been involved at the very end of the process and consequently has been caught in a situation of trying to stop an aspect of the overall watershed management effort. The consequences of not being inclusive? Lawsuits, plans fall through, plans are not implemented, people fail to understand the issue, and all sorts of problems can erupt. Handle concerns early. Build public trust before the partnership asks for action. People support ideas suggested by steering committee members they trust. If public participation is successful, the partnership should not have to fight city halls to have an impact and get support.

The partnerships' public participation efforts should look for opportunities to include interactions between multiple agencies and multiple levels of government. It is difficult to get the attention of communities not directly impacted by a particular problem. The partnership needs to work with adjacent public officials to ensure involvement distant communities.[2] These interactions typically include information exchange, resource sharing, and shared decision making. The existence of interagency communication does not imply any other type of interaction. For organizations to be capable of doing any work together, they must move from the communication stage to the cooperation stage. Cooperation requires some recognition on the part of each organization that there is a common interest best met by some cooperative effort. Figure 4.3 in the Chapter 4 shows the continuum of partnership relationships.

Reassessing assumptions is one source of renewal. Plans can evolve by evaluating feedback about the various outreach practices, initiatives, and efforts. A number of basic steps exist between the starting point — where apathy, a lack of awareness, lack of confidence reign — and some sort of ideal condition where all residents are

Social Capacity Building 221

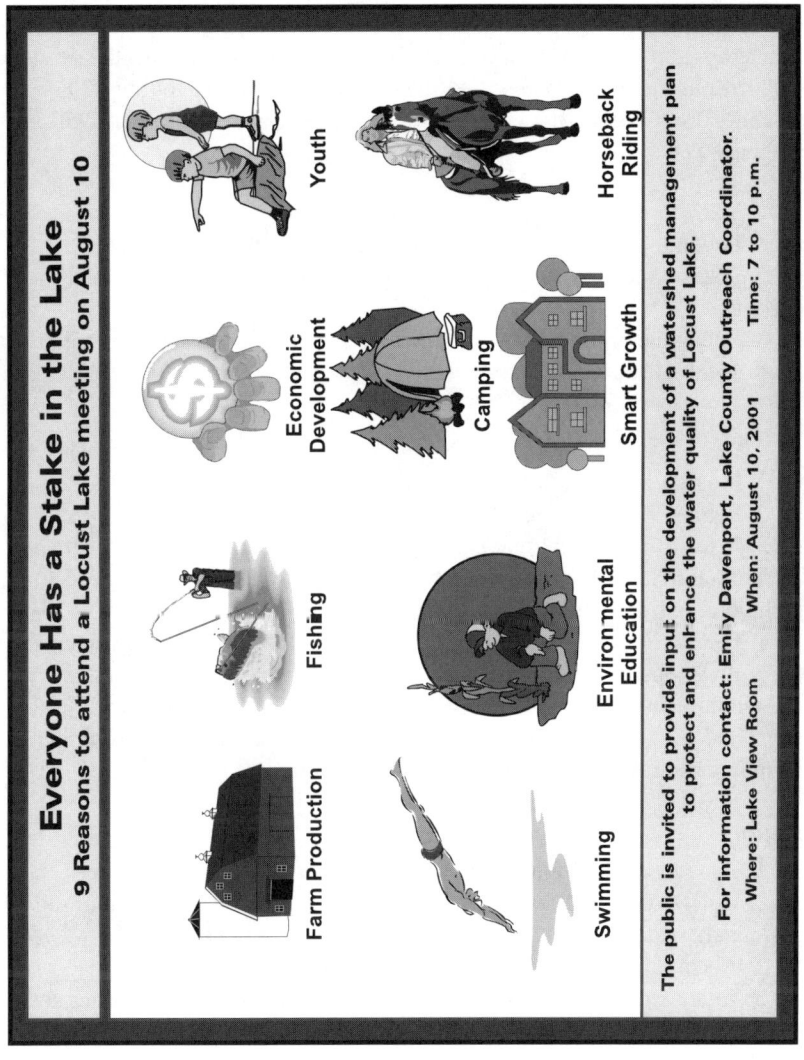

FIGURE 12.4 Newpaper advertisement for community meeting about local watershed management plan.

constructively engaged in the long-term management of the watershed. There are indicators of success for each step along the way. The steps provide a set of benchmarks for people who want to be sure that progress continues steadily. In the real world progress along these lines is never linear. A partnership, for example, may have the right message but use the wrong channel to communicate it.

Public information ranges in type and intensity. While the overall goal is having uniform and active public involvement, not everyone needs to, nor wants to be or can be, fully involved in every decision in the watershed management process. The level, frequency, and purpose of public involvement vary according to the needs of the watershed management process and interests of the various "publics." For different "publics" and needs, at various times in the process, involvement will have different goals.

Awareness

Phase 1. Become informed about the problems and issues.
Phase 2. Focus awareness on proposed solutions and possible outcomes and impacts.
Phase 3. Adjust the process based on the implementation status and potential.
Phase 4. Celebrate success.

Comment

Phase 1. Provide input on the problems and issues.
Phase 2. Review and evaluate alternatives.
Phase 3. A well-informed public is an asset to monitoring and resource management activities. If citizens are aware of the problem and the need for pollution control, then they are likely to support monitoring.
Phase 4. The public will also want to be informed of results in a timely manner. Carefully prepared newspaper articles or press releases are very effective in communicating results. Additional information should be available to interested citizens, and the project coordinator and steering committee members should be available to answer questions.

Participation

Phase 1. Provide information, opinions, and concerns.
Phase 2. Work to create alternatives (providing or analyzing information, suggesting alternative solutions). Enlist community help in planning the watershed project. Hold a community design charrette for their areas.
Phase 3. Volunteer to work on implementing the plan. Organize planting and clean-up days. To stem vandalism, seek to involve kids and youths; they will take care of what they help build.

Action

Phase 1. Collect and provide data to the assessment process.
Phase 2. Help develop solutions, working with various other publics to provide support.

Social Capacity Building

Phase 3. Assume responsibility for actions that will contribute to achieving the goals and objectives outlined in the plan.

12.6 PLANNING FOR SUCCESSFUL PARTICIPATION

A public participation effort is a series of activities that fulfill the public's needs and achieve the goals of the watershed partnership. Public participation often brings to mind public meetings. While meetings are a useful tool for garnering public input, there are many other ways to obtain input from the public, such as interviews, newsletters, design charrettes, or focus groups. When selecting public participation methods, the outreach team needs to consider its commitment to obtain and use public input; expectations and needs of the public; availability of necessary skills, time, and other resources. Public participation activities inform the public, provide opportunities for comment and involvement, respond to input, and provide the forum for people to become involved. These types of activities take time and staff, cost money, and need to be integrated into the overall outreach efforts for the watershed management efforts. The following eight-step planning process is recommended to provide minimum support for public participation. The outcome of each step is also included.

1. Define the reasons for public participation. The product needs to be a written statement of public participation goals. This written statement guides public participation activities and is used to allocate resources to support the effort.
2. Identify the decision points in the planning and implementation processes. The product is an analysis of how decisions will be made in the watershed planning and implementation process. A schedule should be included in the analysis.
3. Analyze the publics. The product is a listing and description of each of the publics and a preliminary recommendation regarding possible public participation roles for each public.
4. Identify what information needs to be provided to the publics. The product is a list of subjects to be discussed and attitudes, concerns, and issues to be included.
5. Define public participation goals and objectives. The product is a written statement of public participation objectives in terms of roles for the identified publics. The objectives directly determine the choice of activities for public participation.
6. Design activities and efforts to achieve public participation objectives. The product is a list of activities and efforts that will accomplish the participation objectives.
7. Prepare public participation component for inclusion in the overall outreach plan. The product is a written plan documenting what is to be done, when, by whom, and for what objective. While this plan is a section of the outreach plan, it has its own discrete budget.
8. Incorporate the results of the public participation into the watershed management process. The product is the documentation of how public partic-

ipation was utilized in the decision-making process. Such documentation is important in evaluating the overall success or failure of the process in achieving the goals outlined in the watershed management plan.

The watershed outreach team should adapt these eight steps to fit its own needs and circumstances. These steps need not occur in the sequence in which they are listed above; several may occur simultaneously, and some steps may need to be repeated as ideas develop and are refined. However, for long-term success in public participation, the activities covered in the eight-step process should occur.

12.7 ANNUAL REVIEW

Since the outreach effort is to be dynamic, an annual update and revisions are recommended. The watershed management plan's outreach strategy represents a major planning effort. Its usefulness will diminish if it is not regularly evaluated and revised to meet current conditions and opportunities. The review covers the last year and sets the stage for the next year's activities. Think in terms of (1) throwing out what is not working, (2) what needs to be strengthened, and (3) what new approaches are needed. The review should involve the outreach team and the steering committee. The review should be scheduled in relationship to work load and budget cycles. Scheduling it for midwinter (January/February), depending on the outreach activities scheduled, provides sufficient time to obtain information on the last year's activities and to analyze it and still be early enough to have an impact on next year's activities. The value of having good evaluation criteria and measurable objectives becomes apparent during the annual review. The next year's outreach action plan is developed to specify activity, lead, cooperators, specific timeframes, costs, and evaluation criteria by objective. The annual action plan is based on the overall outreach strategy, past year's accomplishments, current conditions, and available funding.

Additional resources are:

Getting in Step: A Guide to Effective Outreach in Your Watershed, The Council of State Governments; Washington, D.C.
Public Involvement for Better Decisions, A Guidance Manual, Association of State and Instate Water Pollution Control Administrators, Washington, D.C.
International Association for Public Participation web site: http://www.lap2.org.

REFERENCES

1. Riley, A., *Restoring Streams in Cities, a Guide for Planners, Policymakers, and Citizens*, Island Press, Washington, D.C., 1998.
2. DeShazo, R.P. and Garrigan, P., Merrimack River Initiative Watershed Connections, Lessons Learned in Subwatersheds of the Merrimack River Watershed: The Nashua, Souhegan, and Stony Brook Watersheds, New England Interstate Water Pollution Control Commission, Wilmington, NH, 1996.

3. Simpson, J., Milwaukee survey used to design pollution prevention program, Technical note 37, in *The Practice of Watershed Protection*, T.R. Schueler and H.K. Holland, Eds., Center for Watershed Protection, Ellicot City, MD, 2000.
4. University of Wisconsin Extension (UWEX), UWEX I&E Program Planner for Priority Watershed Projects, UWEX, Madison, 1996.
5. MCOS, Three Rivers Project, Water Quality Monitoring and Management—interim report, MCOS, Dun Laoghaire, Ireland, 1999.
6. USEPA, *Evaluation of Experimental Rural Clean Water Program,* EPA-841-R-93–005, USEPA, Washington, D.C., 1993.
7. Marshall, B., Dealing with polluted runoff in a state scenic river, *River Management Society (RMS) News*, Winter 2001.
8. Holdren, C., Jones, W., and Taggart, J., *Managing Lakes and Reservoirs*, N. Am. Lake Manage. Soc. and Terrene Inst., Madison, 2001.
9. Huron River Watershed Council, Developing a Communications Plan, a Roadmap to Success, Huron River Watershed Council, Ann Arbor, MI, 1996.
10. Dahme, J. and Smullen, J.T., Innovative strategy helps Philadelphia manage combined sewer overflows, *Storm water*, 1(1), 52–57, Nov/Dec 2000.

13 Conclusions

In an ideal world you'd know all the risks and all the benefits before you use something, but we'd be very slow to progress if we had to know all that.

— Dr. Hans S. Aldwinckle

The watershed management process offers a ray of hope in controlling and preventing pollution and restoring water quality. Watershed management based on the four-step process promotes resource outcomes versus the traditional administrative-based management. The goal of the effort switches to attaining designated use instead of program execution. The evaluation measures for the watershed effort are biological, chemical, and physical based not just on administrative actions. The institutional results for agencies involved in watershed management are programs and tools to improve the environment. The administrative actions are evaluated against changes in environmental indicators and conditions rather than improved timeliness and reduced backlogs in administrative actions. Depending on its outcome, the process will either put solutions in place or serve as a stepping stone in the adaptive management cycle to control this omnipresent water-quality issue. Adaptive management involves adjusting management direction as new information becomes available. It requires a willingness on the behalf of the partnership and funding entities to experience failure. Since watershed management efforts are based on the analysis of available information, adequate monitoring is needed. Adaptive management partnerships can focus on and correct specific problems.

The process uniquely finds the balance between economic and social values and needs with the water-quality goals for a watershed. It strikes a compromise where educated stakeholders make the decisions with help from scientists and other professionals. This type of effort is characterized by decentralized and shared decision making, collaboration, engagement of a wide array of stakeholders, and goals evidencing concern for ecosystems. Without a doubt, the success or failure of a watershed management effort depends on three factors: involvement and support of stakeholders and governmental entities, an adequate plan to address the environmental issues, and adequate resources (technical, financial, and outreach) to implement the plan. The important thing is to pull together a partnership that is a manageable size, creates synergy, and represents the key interests in the watershed.

There is a continuing need to monitor, assess, evaluate, and adapt the four-step process. Tribal, state, and federal agency program integration and support for locally based watershed management partnerships represent an ongoing evolution.

A review of successful watershed management efforts indicates several key ingredients are among these projects:

- Community involvement in the planning, implementation, and long-term maintenance of a watershed project is needed to ensure success. Watershed planning begins with a clear understanding of the concerns of the people living in the watershed and the immediate area, as well as the priorities of the involved agencies. In addition, it has been found that community involvement reduced costs to the government agencies since citizens and interest groups donated labor, money, and supplies to a number of restoration projects within watersheds.
- Individuals and organizations provide strong leadership to get the job done. Optimally, a project that has its beginnings and foundation at the local level is likely to achieve its goals.
- A political environment that promotes environmental protection because of the strength and backing of elected officials is key for dealing with funding and land-use issues.
- Interagency participation helps to ensure that the necessary technical skills and resources are available and applied. Interagency participation increases coordination, builds public support, and in most cases broadens public involvement.
- Well-defined goals for a watershed project increase its probability of success. Well-defined goals for the watershed project will help focus implementation activities on achieving those goals. The goals must be realistic and related to what the watershed and its water resources can support, i.e., do not try to make it into something that it is not.
- Management efforts are allotted adequate time and financial support without which the efforts will fail. The adequate time provision needs to include time after the majority of the implementation has been completed to ensure the techniques and practices function properly. In cases where vegetative practices are used, a time period for adequate establishment needs to be considered.
- Management approaches that promote systems that protect the environment and increase participants and profitability, such as CORE4, are a key to acceptance by rural residents.
- Public awareness of proper land-use practices is important for the long-term preservation and conservation in the watershed. Where the public and individual landowners recognize the impacts of poor land-use practices, they are likely to support management and restoration activities.
- Demonstration projects are useful tools for building acceptance and support for the large-scale implementation within a watershed. Demonstration projects should also include maintenance and operation aspects.
- Postimplementation monitoring is needed to document the overall impact of the watershed project to ensure no unforeseen adverse impacts to the downstream community emerge.

Conclusions

- A comprehensive watershed approach is used in transboundary issues because, while site-specific and localized, the problem is caused by upstream management within another jurisdiction. Another need is to ensure that the solution to a water-quality problem does not create an additional downstream problem.

Remember:

1. Assemble the team.
   Identify and bring together partners.
   Discuss and develop a common vision for a healthy watershed.
2. Look at the big picture.
   Assess the scope of concerns in the watershed.
   Collect information/data from all partners and other sources.
3. Analyze data.
   Look at small and large problems.
   Look at all areas of the watershed.
   Identify challenges and opportunities.
4. Focus on manageable tasks.
   Identify critical natural resource concerns and geographic areas.
   Develop goals and objectives for those prioritized concerns/areas.
   Formulate and select alternatives for reaching those goals.
   Consider how to perform actions, including the daily/monthly maintenance requirements that will keep the system functioning.
5. Implement.
   Create a master schedule for prioritized actions.
   Assign responsibilities.
   Dedicate funds for the tasks.
   Complete the task.
   Know when and what to evaluate.
   Document the results.
6. Review and revise.
   Schedule 6- and 12-month reviews of progress.
   Ask the key questions.
   Check against projected outcomes.
   Revise actions/responsibilities/budgets as necessary.
7. Report.
   Success and failures are both building blocks to the future.

Good luck!

Glossary

acid deposition a complex chemical and atmospheric phenomenon that occurs when sulfur and nitrogen compounds and other substances are transformed by chemical processes in the atmosphere; this transformation is often deposited on Earth in either wet or dry form; the wet forms are often called acid rain; the dry forms are acidic gases or particulates

acute effect occurs in response to short-term exposure to a pollutant

acute pollution (nonpoint source) episodic or randomly occurring inputs of pollution to a waterbody; examples include agricultural runoff that enters a receiving water during storm events; may have a long-lasting impact to the aquatic system

adsorbed attached to the surface of the water

adsorption the process by which pesticides and other chemicals are attracted to the surface of a soil or organic particle

aerobic environmental conditions characterized by presence of dissolved oxygen; used to describe biological or chemical processes that occur in the presence of oxygen

aggradation raising the bed of a watercourse by deposition of sediment

aggregate mass or cluster of soil particles

agricultural pollution liquid, dissolved, and solid wastes from all types of farming; including runoff from pesticides, fertilizers and feedlots, erosion and dust from plowing, animal manure and carcasses, crop residue and debris

agricultural runoff portion of precipitation on an agricultural drainage area that does not infiltrate into the ground; the drainage area may include areas of crop production, pastures, rangeland, or feedlots

algae any organisms of a group of chiefly aquatic microscopic nonvascular plants; most algae have a chlorophyll as the primary pigment for carbon fixation; as primary producers, algae serve as the base of the aquatic food chain; an overabundance of algae in natural waters is known as eutrophication

algal bloom rapidly occurring growth and accumulation of algae within a waterbody; it usually results from excessive nutrient loading and sluggish circulation regime with a long residence time; persistent and frequent blooms can result in low oxygen conditions

allocations that portion of a receiving water's loading capacity attributed to one of its existing or future pollution sources (nonpoint or point) or to natural background sources

alluvium sediment deposited by flowing water, such as in a riverbed, floodplain, or delta

alternative strategies a set of one or more strategies developed during the planning process to solve resource problems identified during the assessment and problem identification phase to achieve proper management of the resources of concern

ambient monitoring water-quality tests taken at specific sites along a watercourse or within an aquifer at specific times over an extended period to determine long-term changes in the water system; all forms of monitoring conducted beyond the immediate influence of a discharge pipe or injection well

ambient water quality natural concentration of water-quality constituents prior to mixing of either point or nonpoint source load of contaminants

anaerobic environmental condition characterized by zero oxygen levels; describes biological and chemical processes that occur in the absence of oxygen

anoxia a condition of no oxygen on the water

anoxic aquatic environmental condition characterized by zero or little dissolved oxygen

anthropogenic eutrophication the accelerated aging of a lake as a result of human activities; normally due to nutrient and sediment loading

aquatic corridor areas of land and water important to the integrity and quality of a stream or other body of water; an aquatic corridor usually consists of the actual stream, the aquatic buffer, and other areas that are a part of the stream's right-of-way

aquifer underground layers of porous rock saturated with water

assessment the translation of scientific data into policy-relevant information suitable for supporting decision making and action

assimilative capacity the amount of contaminant load that can be discharged to a specific waterbody without exceeding water-quality standards or criteria. Assimilative capacity is used to define the ability of a waterbody to naturally absorb and use a discharged substance without impairing water quality or harming aquatic life

atmospheric deposition the accretion of chemicals including nitrogen and phosphorus attached to dust materials during dry weather or as part of raindrops during wet weather, which deposit onto the land or water surfaces from the air

background levels levels representing the chemical, physical, and biological conditions that would result from natural geomorphological processes such as weathering or dissolution

bacteria a large diverse group of single-celled microscopic organisms; some are important in agriculture, some are important in the decay of organic matter

base fixed sites sites on streams at which stream flow is measured and samples are collected to assess the broad-scale spatial and temporal character and transport of inorganic constituents of stream water in relation to hydrologic conditions and environmental settings

base flow the flow of a stream in dry weather, which comes from a watershed's groundwater; sometimes called seasonal low flow

Glossary

baseline initial or existing conditions of flux before management intervention or impact

basin the largest single watershed management unit for water planning, that combines the drainage of a series of subbasins; often have a total area of more than a thousand square miles

BASINS (better assessment science integrating point and nonpoint sources) a computer-based tool that contains an assessment and planning component that allows users to organize and display geographic information for selected watersheds

BAT best available technology; the best available technology treatment techniques, or other means which the regulatory agency finds, after examination for efficacy under filed conditions and not solely under laboratory conditions, are available (taking cost into consideration)

bathymetric measurements of lake basin, such as water depth, sediment depth, relief of bottom, or volume

benchmark a surveyor's mark made on a stationary object of previously determined position and elevation and used as a reference point in surveys

beneficial use the use of water resources embodied within the Clean Water Act; may also be the most beneficial use of the resource by the community within which it lies

benthic community life on the bottom of waterbodies or in the sediments of a waterbody

best management practices measures (structural, vegetative, and management) that are the most effective and practical means of preventing or reducing pollution inputs from nonpoint sources to waterbodies

biennially once every 2 years

bioaccumulation the process by which organisms absorb and retain toxic substances from their environment

bioassessment biological assessment; the evaluation of an ecosystem using integrated assessment of habitat and biological communities in comparison to empirically defined reference conditions

biochemical oxygen demand (BOD) a measure of the quantity of dissolved oxygen removed from water by the metabolism of microorganisms; excessive BOD results in oxygen-poor water

biodiversity a variety of natural plant, aquatic, and animal communities within the watershed

biofiltration the use of a series of vegetated swales to provide filtering treatment for stormwater as it is conveyed through the channel

biological criteria (biocriteria) quantitative and narrative goals for the aquatic community used within the water programs to refine use designations, establish criteria for determining use attainment/nonattainment, evaluate effectiveness of current water programs, and detect and characterize previously unknown impairments

biological monitoring the use of aquatic biota entity as a detector and its response as a measure to determine environmental conditions

biota the animals and plants that live in a particular location or regions

blue-green algae algal form that may cause water to turn green, gray, or brown during late summer periods; some forms may be toxic in large concentrations

brownfield a site that was previously used for industrial or other purposes that may have contaminated the soils there

buffer strip or zone strips of erosion-resistant vegetation between a waterway and an area of more intensive land use

calibration (model) the process of adjusting model parameters within physically defensible ranges until the resulting predictions give a best possible good fit to observed data

chain-of-custody a method of tracking collection and analysis procedures, including each person who handles a sample; usually outlined in QAPP

channel a natural stream that conveys water; a ditch or open manmade conveyance for the flow of water

channelization deepening, widening, and straightening a channel of a stream to increase its water-carrying capacity; the activity usually results in the loss of riparian vegetation

chronic pollution (point) pollution inputs marked by a long duration and frequent recurrence; examples include the municipal or industrial wastewater discharges

Clean Water Act (CWA) the Clean Water Act, PL 92–500, as amended by PL 96–483 and PL 97–117, contains a number of provisions to restore and maintain the quality of the nation's water resources

coastal zone land and water adjacent to the coast that exerts an influence on the uses of the ocean and its ecology or whose uses and ecology are affected by the ocean

combined sewer a sewer receiving both intercepted surface runoff and municipal sewage

community (ecological) an aggregate of organisms that form a distinct ecological unit

compliance the act of fulfilling an official requirement; submission to operative laws, regulations, terms, and conditions

concentration mass per unit volume such as milligrams per liter

confluence point where two or more watercourses intersect

consensus a process that results in a decision that everyone can live with and everyone agrees to support and work toward

conservation plans formal plans to protect natural resources

conservation targets quantifiable objectives a planning committee can use in prioritizing activities and evaluating success of its efforts

conservation tillage any tillage and planting system that maintains at least 30% of the soil surface covered by residue after planting to reduce soil erosion by water; where soil erosion by wind is the primary concern; maintains at least 1000 lbs of small grain residue equivalent on the surface during the critical erosion period

contaminant any physical, chemical, biological, or radiological substance or matter that has an adverse effect on air, water, or soil

cost–benefit analysis a quantitative evaluation of the costs that would be incurred versus the overall benefits to society of a proposed action such as the establishment of acceptable loading of a pollutant of concern

cost sharing a publicly financed program through which society, the beneficiary of environment protection, shares part of the cost of pollution control with those who must actually install the controls

critical area the part of the watershed contributing a majority of the pollutants and having the most significant impact on the waterbody

critical condition the critical condition can be thought of as the "worst-case" scenario of environmental conditions in the waterbody in which the loading expressed in the TMDL for the pollutant of concern will continue to meet water-quality standards; critical conditions are a combination of environmental factors that results in attaining and maintaining the water-quality criterion

crop rotation growing different crops in recurring succession on the same land, as opposed to continuous culture of one crop

current practice how practitioners working in the field of watershed management apply the science

data quality the totality of features and characteristics of data that defines its ability to satisfy a given purpose; the characteristics of major importance are accuracy, precision, completeness, representation, and comparability

data validation a systematic process for reviewing a body of data against a set of criteria to provide assurance that the data are adequate for their intended use; data validation consists of data editing, screening, checking, auditing, verification, certification, and review

degradation a noticeable decline in the health of a natural waterbody

design charrette an intensive workshop to develop objectives and design ideas

designated uses uses specified in water-quality standards for each waterbody or segment thereof, whether or not they are attained; typical uses are navigation, hydro-electric power generation, drinking-water supply, fish propagation, and recreation (swimming, boating, fishing, etc.)

detachment the process by which a substance becomes mobilized or available for transport; one of three stages of the pollutant delivery triangle

detention the slowing of flows — either entering the sewer system or draining over the surface — by temporarily holding the water on a surface area in a storage basin

deterministic model a model that does not include built-in variability; same input will always equal the same output

direct discharge also known as point source emissions, direct discharge refers to any intentional release of wastes through direct dumping or pipeline discharge

direct runoff water that flows over the ground surface or through the ground directly into streams, rivers, and lakes

discharge volume of water per unit time moving past a fixed point

discharge monitoring report (DMR) report of effluent characteristics submitted by a permittee under an NPDES discharge permit

discharger a facility that releases effluent into a waterbody from a pipe, or other point source mechanism requiring an NPDES, state, or tribal permit

dissolved oxygen (DO) atmospheric oxygen dissolved in water or wastewater, commonly employed as a measure of water quality; low levels adversely affect aquatic life

diversion individually designed conveyances across a hillside; may be used to protect bottom land from hillside runoff, divert water from aerial sources of pollution, or protect structures from runoff

domestic wastewater also called sanitary wastewater; consists of wastewater discharged from residence and from commercial, institutional, or similar facilities

drainage the removal of excess water from land by means of ditching or subsurface infiltration

drainage basin a part of a land area enclosed by a topographic divide from which direct surface runoff from precipitation normally drains by gravity into a receiving water

drainage density the ratio of the total length of streams within a watershed to the total area of the watershed; thus drainage density has units of the reciprocal of length; a high value of the drainage density would indicate a relatively high density of streams and thus a rapid storm response

drainage tile pipe installed for internal drainage purposes

drainage pattern the configuration of arrangement in plan view of the natural stream courses in a watershed; it is related to local geologic and geomorphologic features and history

dynamic model a mathematical formulation describing and simulating the physical behavior of a system or a process and its temporal variability

dynamic simulation modeling the behavior of physical, chemical, and biological phenomena and their variation over time

easement a limited right someone else holds over land

ecological conditions the degree of functionality or health of an ecosystem, measured by a broad array of indicators of condition that includes biotic characteristics and abiotic characteristics

ecological integrity a measure of the health of the entire area or community based on how much of the original physical, biological, and chemical components of the area remains intact

ecological synoptic study study of biological communities and habitat characteristics to evaluate the effects of physical and chemical characteristics of water and hydrologic conditions on aquatic biota and habitat characteristic of a geographic area

ecoregion a physical region defined by its ecology, based on similar soils, land surface, natural vegetation, and current land use

ecosystem a system of interrelated of animals, plants, and the physical–chemical environment within which they function and interact

effluent the treated or untreated liquids that flow out or are discharged from a water treatment plant, sewer, or industrial outfall

empirical model use of statistical techniques to discern patterns or relationships underlying observed or measured data for large sample sets; does not account for physical dynamics of waterbodies

enhancement in the context of restoration ecology, any improvement of a structural or functional attribute

environmental framework natural and human-related features of the land and hydrologic system, such as geology, land use, and habitat, that provide a unifying framework for making comparative assessments of the factors that govern water-quality conditions within and among watersheds

environmental indicators measurable features that can be used to determine if the narrative or numeric management goals are being met

environmental setting land areas characterized by a unique, homogeneous combination of natural and human-related factors, such as row-crop cultivation on glacial-tills

ephemeral refers to flow that dries up as some time during the year, usually summer

erosion the wearing away of the land surface by running water, wind, ice, or other geological agents including such processes as gravitational creep

estuary region of interaction between rivers and near-shore ocean waters; tidal action and river flows mix fresh and salt water; such areas include bays, mouths of rivers, salt marshes, and lagoons; these brackish water ecosystems shelter and feed marine life, birds, and wildlife

eutrophic a nutrient-rich trophic condition

eutrophication the process of enrichment of waterbodies by nutrients; eutrophication of a lake normally contributes to its slow evolution into a wetland and ultimately to dry land; may be accelerated by human activities

explanatory variable a statistical term for "variable" that helps explain the variability in the dependent term

export mass of pollutant lost from unit area per unit time

fecal coliform bacteria present in the intestinal tract of warm-blooded animals; indicators that disease-causing bacteria or viruses may be present

fecal streptococcus a group of bacteria normally present in large numbers in the intestinal tracts of warm-blooded animals such as human beings and domesticated animals

feedlot a lot or building or a groups of lots or buildings intended for confined feeding, breeding, raising, or holding animals

fertilizer any organic or inorganic material of natural or synthetic origin added to soil to provide one or more plant nutrients

fertilizer management application of fertilizers consistent with soil and crop needs without loss with runoff water through deep percolation or means; a practice designed to minimize the contamination of surface and groundwater by limiting the amount of nutrients applied to soil to no lore than the crop is expected to use

floodplain a nearly flat area of land along the course of a stream that is naturally subject to flooding

fluvial pertaining to or produced by action of a stream; or existing, growing, or living in or near a stream

fluvial geomorphology the effect of rainfall and runoff on the form and pattern of riverbeds and river channels

flux the rate at which a measurable amount of material flows past a designated point in a given amount of time

food chain a sequence of organisms, where each uses the next-lower member of the sequence as a food source

food web the totality of interacting food chains in an ecological community

forebay additional storage space located near a stormwater practice inlet that serves to trap incoming coarse sediments before they accumulate in the main treatment area

full build-out the total potential development in a watershed based on current zoning plans which includes the existing development and expansion potential in the future

geographic information system (GIS) a system that integrates layers of spatially oriented information either manually or automatically

geomorphology the study of the evolution and configuration of landforms

global positioning system (GPS) a system capable of providing worldwide navigation and positioning by pinpointing locations

grab sample a single sample of water collected at an arbitrary time and flow

grassed swale an earthen conveyance system in which the filtering action of grass and soil infiltration are utilized to remove pollutants from urban and agricultural stormwater: an enhanced grass swale, or biofilter, utilizes check dams and wide depressions to increase runoff storage and promote greater settling of pollutants

grassed waterway a natural or constructed waterway used to conduct surface water from or through cropland

groundwater water that accumulates in the spaces between alluvial material (sand, gravel, silt, or clay) or in fractures of rocks

habitat the area where a plant or animal naturally lives

habitat modification loss of habitat that occurs when streams or rivers are flooded and when streams, rivers, and lakes are modified by activities such as grazing, farming, channelization, construction of dams, and dredging; typical examples of habitat modification include loss of stream-side vegetation, siltation, smothering of bottom-dwelling organisms, and increased water temperatures

headwaters the origin and upper reaches of a river or stream

heavy metals metals with a high molecular weight like mercury, cadmium, arsenic, lead, copper, chromium, and zinc, that poison living organisms at low concentrations and tend to accumulate in the food chain

herbicide a chemical substance used for killing plants, especially weeds

holistic emphasizing the importance of the whole and the interdependence of its parts

hydric soil a soil that is saturated, flooded, or ponded long enough to support the growth of wetlands vegetation

Glossary 239

hydrodynamic model mathematical formulation used in describing fluid flow of circulation, transport, and deposition processes in receiving water

hydrologic cycle the circuit of water movement from the atmosphere to earth and its return to the atmosphere through various stages or processes, such as precipitation, interception, runoff, infiltration, storage, evaporation, and transpiration

hydrologic modification any change in the natural stream configuration such as channelization or dredging

hydrologically distinct defined by drainage basins or watersheds rather than arbitrarily defined by political boundaries

hydrology the study of occurrence distribution and chemistry of all waters on or below the Earth's surface and in the atmosphere

hydromodifcation changing the flow, and thereby habitats, of natural water systems; this process includes the construction of dams, stream channels, and canals

hypoxia the terms "hypoxia" and "hypoxic waters" refer to waters with concentrations of less than two parts per million of dissolved oxygen, which is generally accepted as the minimum level required to support most animal life and reproduction

impaired degradation of water quality below state standards for a waterbody's beneficial uses

impairment a detrimental effect on the biological integrity of a waterbody caused by an impact that prevents attainment of the designated use

impervious cover any surface in the urban landscape that cannot effectively absorb or infiltrate rainfall or runoff

imperviousness the percentage of impervious cover within a development site or watershed

implementation monitoring (also known as tracking) documents whether or not management practices were applied as proposed; project administration is part of implementation monitoring

implementation plan an outline of steps and activities needed to meet environmental quality standards by a set time

Index of Biological Integrity (IBI) tool for assessing the effects of runoff on the quality of the aquatic ecosystem by comparing the condition of multiple groups of organisms or taxa against the levels one expects to find in a healthy stream

indicator a measurable quantity that can be used to evaluate the relationship between pollutant sources and their impact on water quality

indicator sites stream sampling sites located at the outlets of a subwatershed or watershed with homogeneous land use and physiographic conditions

individual systems see septic systems

infiltration the gradual downward movement of water from the surface into the subsoil

infiltration capacity the capacity of soil to allow water to infiltrate into or through it during a storm

information networks the web of personal relationships that individuals draw upon to gain access to information, resources, and assistance

inorganic waste waste material such as sand, salt, iron, calcium, and other minerals that are not converted in large quantities by biological activity

integrated pest management (IPM) a mixture of chemical and other, nonpesticide, methods to control pests

interflow the lateral flow of water through soil

intermittent stream flow at certain times of year

interstate water waterbody located in two or more states and tribes

inventory the collection of natural resource, economic, and social information within the watershed

irrigation return flow surface and subsurface water that leaves a field after the application of irrigation water

Karst the type of geologic terrain underlain by carbonate; rocks; occuring due to flowing groundwater

Karst topography land area characterized by depressions without external drainage

knowledge base refers to how widely scientific understanding is shared within the scientific community

lake a considerable inland body of standing water; either naturally formed or manmade

land application discharge of wastewater and related solids, manure, or sewage sludge onto the ground for treatment or reuse

leachate water that collects contaminants as it trickles through wastes, pesticides, or fertilizers

load or loading an amount of matter or thermal energy introduced into a receiving waterbody; to introduce matter or thermal energy into a receiving water

load allocation the portion of receiving water's loading capacity attributed either to one of its existing or future nonpoint sources of pollution or to natural background sources

loading capacity the greatest amount of loading a water can receive without violating water-quality standards

macroinvertebrate invertebrate are the aquatic insects that spend a portion of their life cycle in the water, usually among the rocks and sediment deposits on the bottom

management measures best practical and economically achievable measures to control the addition of pollutants to waterbodies through the application of nonpoint pollution control practices, technologies, processes, siting criteria, operating methods, best management practices, or other alternatives

manure generally the refuse from stables and barnyards including both animal excreta and straw or other litter

mass wasting downslope transport of soil and rocks due to gravitational stress

mathematical model a system of mathematical expressions that describe the spatial and temporal distribution of water-quality constituents resulting from fluid transport and the one, or more, individual processes and interactions within some prototype aquatic ecosystem; a mathematical water-quality model is used as the basis for waste-load allocation evaluations

mesotrophic a trophic condition characterized by intermediate nutrient availability

meta-data summaries information that describes the content, quality, condition, and other characteristics of data

metrics specialized biological variables that can be combined with a rating and used in an index

mitigation actions taken to avoid, reduce, or compensate for the effects of environmental damage; among the broad spectrum of possible actions are those that restore, enhance, create, or replace damaged ecosystems

mixing zones limited areas where initial dilution of discharge takes place; water-quality changes may occur; and certain water-quality standards may be exceeded

monitoring periodic or continuous surveillance or sampling using consistent methods to determine the level of pollution or radioactivity

morphometry measurements of the physical structure (e.g., length of streams, slope, depth, shoreline length) of a watershed or waterbody

National Pollutant Discharge Elimination System (NPDES) established by Section 402 of the Clean Water Act, this federally mandated system is used for regulating point source discharges under Sections 307, 318, 402, and 405; this program is for issuing, modifying, revoking and reissuing, terminating, monitoring, and enforcing permits

natural waters flowing water within a physical system that has developed without human intervention, in which natural processes continue to take place

navigable waters waters sufficiently deep and wide for navigation by all, or specified, vessels; such waters in the United States come under federal jurisdiction and are protected by certain provisions of the Clean Water Act

nonpoint source pollution nonpoint source (NPS) is so named because the pollutants do not originate at a single point sources such as industrial and municipal waste discharge pipes; also known as people or diffuse pollution; in general, nonpoint source pollution is a compilation of land runoff, precipitation, percolation, and atmospheric deposition; while some NPS pollution occurs naturally, most NPS problems are the result of inappropriate land use or management

nonstructural methods nonphysical approaches to pollution controls such as land-use controls, construction activity schedules, and zoning ordinances

N/P ratio the ratio of nitrogen to phosphorus in an aquatic system; the ratio is used as an indicator of the nutrient limiting conditions for algal growth; also used as an indicator for the analysis of trophic levels of receiving water

nuisance species nonnative populations of fish and shellfish that dramatically increase, displacing native species, reducing biodiversity, and limiting water-use activities

numeric target a measurable value determined for the pollutant of concern, which, if achieved, is expected to result in the attainment of water-quality standards in the waterbody of concern

numerical model model that approximates a solution of governing partial differential equations that describe a natural process; the approximation uses a numerical discretization of space and time components of the system or process

nutrients elements or substances, such as nitrogen or phosphorus, that are necessary for plant growth; large amounts of these substances reaching waterbodies can become a nuisance by promoting excessive aquatic algae growth

oligotrophic a nutrient-poor condition

one-dimensional model (1D) a mathematical model defined along one spatial coordinate of a natural water system; typically used to described the longitudinal variation of a watercourse

organic pollutants one of the general groups of pollutants toxic to aquatic life; includes pesticides, solvents, PCBs, dioxins, etc.

organic waste waste material that comes from animal or vegetable sources; generally can be consumed by bacteria and small organisms; contains mainly carbon and hydrogen

outfall the point of discharge for a river, tile drain, conduit, or pipe

overland runoff portion of precipitation that flows from a drainage area on the land surface or in open channels

partnership an association of persons joined in an undertaking as shares or partakers; what the partnership is called and how it is organized is almost entirely up to the group that forms to address the issue; there are two major types of partnerships: operational (coordination issues are not related to a specific geographic area) and planning (partnerships are related to a specific geographic area of concern); the geographic focus watershed partnerships are considered the planning type

pasture an area of land covered with grass or other herbaceous forage plants kept for grazing animals

pathogens disease-causing organisms, especially viruses, bacteria, or fungi

peak runoff the highest value of the stage or discharge attained by a flood or storm event; also referred to as flood peak or peak discharge

perennial refers to flow that occurs throughout the year

permit an authorization, license, or equivalent control document issued by a governing authority to implement the requirements of an environmental regulation

Permit Compliance System (PCS) computerized management information system that contains data on NPDES-permitted facilities

pesticide any chemical agent used to control specific organisms; such as insecticides, herbicides, fungicides

point source specific sources of pollutants often discharged from pipes and covered by site-specific state and federal permit programs

pollutant any substance or material in such quantities that it adversely affects or threatens a resource's usefulness

pollutant delivery triangle summarized pollutant generation process consisting of the following steps: availability, detachment, and transport

pollution the manmade or human-induced alteration of chemical, physical, biological, and radiological integrity of water

pool portion of a stream with reduced current velocity, often with deeper water than surrounding areas and with a smooth surface

precipitation a water deposit on earth in the form of hail, rain, sleet, or snow

preteatment the treatment of wastewater to remove or reduce contaminants prior to discharge into another treatment system or a receiving water

primary treatment the first major stage I wastewater treatment screens and a sedimentation tank are used to remove most materials that float or will settle; primary treatment removes about 35% of carbonaceous biochemical oxygen demand and less than half of the metals and toxic organic substances from domestic sewage

publicly owned treatment works (POTW) any device or system owned by a state or municipality that is used in the treatment of municipal sewage or industrial wastes of a liquid nature

quality assurance/quality control (QA/QC) a system of procedures, checks, audits, and corrective actions to ensure that all research design and performance, environmental monitoring and sampling, and other technical and reporting activities are of the highest quality

rapid bioassessment several protocols developed by the USEPA and several states to examine the biological community of a stream

rapid stream assessment technique a set of protocols developed to provide a simple, quick field-level assessment of stream quality conditions

reach a section of a stream

receiving waters waters of a watercourse or waterbody that receive treated or untreated wastewater or nonpoint source runoff

redevelopment new development activities on previously developed land

remediation action taken to correct a pollution problem

reservoir lake created by artificially damming a stream or river where water is collected and kept in quantity for a variety of uses, including flood control, water supply, and hydroelectric power

resource management system (RMS) a prescribed combination of conservation practices and management identified by land or water uses that, when implemented, prevents resource degradation and permits sustained use of the resource while maintaining acceptable water quality and acceptable ecological and management levels for the selected resource

response indicator an environmental indicator measured to provide evidence of the biological condition of a resource at the organism, population, community, or ecosystem level

restoration return of an ecosystem to close approximation of its presumed condition prior to disturbance

retention the prevention of runoff from entering the drainage system by storing it on a surface area or in a storage basin

retrofit the installation of a new stormwater practice or the improvement of an existing one in a previously developed area

riffle a rocky shoal or sand bar located just below the surface of the water

riparian areas the land area that borders streams, wetlands, lakes, and rivers and that directly affects and is affected by the water quality; this area often coincides with the maximum water surface elevation of the 100-year storm

riparian zone the border or banks of a stream; although this term is used interchangeably with "floodplain," the riparian zone is generally regarded as relatively narrow compared to a floodplain

river delta system habitats located at the point a river empties into a larger body of lotic water; these areas are usually rich in nutrients

runoff the portion of rainfall, melted snow, or irrigation water that flows across the surface or through underground zones and eventually into streams; runoff has three components: surface runoff, interflow, and groundwater flow; runoff may pick up pollutants from the air or the land and carry them to receiving waters

sanitary sewer overflows (SSOs) a sewer that transports only wastewater from residences and industries to a wastewater treatment plant; wastewater and runoff in these sewers may occur in excess of the sewer capacity and cannot be treated immediately; the excess is frequently discharged directly to a receiving stream without treatment or to a holding basin for subsequent treatment and disposal

science defined as scientific understanding

scoping modeling a method of approximation that involves simple, steady-state analytical solutions for a rough analysis of the problem

secondary sources pollutant sources dispersed throughout the watershed whose magnitude cannot easily be estimated from readily available land-use information

secondary treatment the second stage in publicly owned wastewater treatment systems in which bacteria consume the organic parts of the waste; the treatment removes floating solids, solids that can set, and about 90% of the oxygen-demanding substances and suspended solids; disinfection is the final stage of secondary treatment ediment solid material that is in suspension, is being transported, or has been moved from its original location by air, water, gravity, or ice

sedimentation a broad term that embodies the process of particle erosion, transportation, and deposition by flowing water and wind.

seepage percolation of water through the soil from unlined canals, ditches, lateral watercourses, or waste storage facilities

septic system an on-site system designed to treat and dispose of domestic sewage; a typical septic system consists of a tank that receives waste from a residence or business and a system of tile lines or a pit for disposal of the

Glossary

liquid effluent that remains after decomposition of solids by bacteria in the tank; must be pumped out periodically

shoreland area area within 300 feet of a stream or river or 1000 feet of a lake or wetland, or extent of the floodplain

siltation particles carried in water that are deposited on the bottom of a waterbody

silviculture the art and science of controlling the establishment, growth, composition, health, and quality of forests and woodlands to meet the diverse needs and values of landowners and society on a sustainable basis

simulation the process that mimics some or all of the behavior of one system with a different, dissimilar system, particularly with models

spatially referenced data assigning specific geographic locations to data

stakeholder any person, agency, or organization with a stake or interest in the watershed management plan; individuals who live, work, or play in the watershed; in addition to businesses operating in the watershed or relying on resources from watershed, civic, social, conservation, and environment associations and organizations are recognized as stakeholders; local, state, and federal organizations and governments are also considered stakeholders

steady-state model mathematical model of fate and transport that uses constant values of input variables to predict constant values of receiving water-quality concentrations

storm sewer a sewer that carries only surface runoff, street wash, and snow melt from the land; storm sewers are completely separated from those that carry domestic, industrial, and commercial wastewater

stormwater runoff from a storm, snow melt runoff, and surface runoff and drainage

stratification formation of water layers each with specific physical, chemical, and biological characteristics; as density of water decreases due to surface heating, a stable situation develops with lighter water overlaying heavier and denser water

stream order a method of classifying streams according to their relative position in the stream network; a stream with no tributaries is considered a first-order stream; when two first-order streams combine together, they form a second-order stream, and so on

stressors factors that threaten the well-being or health and long-term viability of an ecosystem, species, or a population

structural management practices construction of physical entities for delaying, blocking, or trapping pollutants

substrate the material making up the bed or bottom of a stream or other waterbody

subwatershed a smaller geographic section of a larger watershed unit with a drainage area between 2 to 15 square miles and whose boundaries include all the land area draining to a point where two second-order streams combine to form a third-order stream

surface runoff precipitation excess not retained on the vegetation or surface depressions and not lost by infiltration and thereby is collected on the surface and runs off

surface water all water naturally open to the atmosphere and all springs and wells directly influenced by surface water

suspended solids or load organic and inorganic particles (sediment) suspended in and carried by fluid (water); suspended sediment usually consists of particles <0.1 mm, although size may vary according to current hydrological conditions; particles between 0.1 mm and 1.0 mm may move as suspended or be deposited (bedload)

sustainability meeting the needs of the present without compromising the ability of future generations to meet their own needs

technology-based controls uniform national requirements for the pollution controls that industrial and municipal dischargers must achieve; technology-based effluent limitations are developed on an industry-by-industry basis

terraces an embankment, or combination of embankment and channel, constructed across a slope to control erosion by reducing the slope and by diverting or storing surface runoff instead of permitting it to flow uninterrupted down the slope

tertiary treatment the advance cleaning of wastewater that goes beyond the secondary stage using such processes as nutrient removal, filtration, or carbon adsorption; the step typically achieves about 95% removal of nutrients, biological demand, and suspended solids

thermal pollution the discharge of water sufficiently warm to lower dissolved oxygen levels, cause eutrophication, affect the life processes of aquatic organisms, or degrade the quality of drinking water or recreational use

threatened waters waters presently meeting state criteria for beneficial use classification, but conditions in the watershed could lead to impairment if control of nonpoint source pollution is not implemented

three-dimensional model mathematical model defined along three spatial coordinates where the water-quality constituents are considered to vary over all three spatial coordinates of length, width, and depth

topographic map land map that display elevations along with natural and manmade features

topography the physical features of a geographic surface area including relative elevations and the positions of natural and manmade features

total maximum daily load (TMDL) a calculation of the maximum amount of pollutant (constituent) that a waterbody can receive and still meet water-quality standards, and an allocation of that amount to the pollutant's sources; also refers to the process of estimating this quantity and allocating it among point and NPSs

transport the process by which a pollutant moves from one location to another

tributary a stream or river that joins and feeds into a larger stream, river, or other body of water

trophic state a classification of the condition of a waterbody pertaining to the degree of eutrophication of a lake; trophic states include oligotrophy (nutrient-poor), mesotrophy (intermediate nutrient availability), eutrophy (nutrient-rich), and hypertrophy (excessive nutrient availability)

turbidity a measure of water clarity (opacity) caused by suspended sediments and organic materails in water

two-dimensional model a mathematical model defined along two spatial coordinates where the water-quality properties are considered averaged over the third remaining spatial coordinate

upstream waters rivers, creeks, and tributaries that empty to another; also, any water located in the opposite direction of the current of a river, creek, or other tributary

urban runoff stormwater from city streets, usually carrying litter, sediment, petroleum products, lawn fertilizers and pesticides, road deicers, and organic wastes

use attainability analysis (UAA) a type of beneficial use analysis that is a structured scientific, multifaceted assessment of the physical, chemical, biological, and economic factors that affect the attainment of the use of the water resource

use attainment the degree to which the water resource meets its designated use

validation the process of determining how well the mathematical model's computer representation describes the actual behavior of the physical processes under investigation; a validated model will have also been tested to ascertain whether it accurately and correctly solves the equations used to define the system simulation

variable term used to describe a quantity that has no fixed value

vegetative controls control measures or practices that involve plants to reduce erosion and treat runoff

vegetative strips grasses or legumes planted in collection ditches to help trap sediments and protect the ditch from erosion

verification (of a model) testing the accuracy and predictive capabilities of a calibrated model on a data set independent of the data set used for calibration

waterbody a geographically defined portion of navigable waters, made up of one or more of the segments of rivers, streams, lakes, wetlands, and coastal waters

water pollution any introduction of foreign material into water or other impingement upon water that produces undesirable changes in the physical, biological, or chemical characteristics of the water

water quality the biological, chemical, and physical conditions of a waterbody, often measured by its ability to support life

water-quality assessment the determination of whether a waterbody is attaining its designated uses for such purposes as drinking, contact recreation, fisheries and irrigation, based on water-quality standards as provided for in the Clean Water Act of 1987

water-quality–based limits or controls water-quality–based controls are more stringent than technology-based controls and are used when technology-based controls are inadequate to achieve water-quality standards

water-quality standards minimum requirements of purity of water for various designated uses; the three components of water-quality standards include the beneficial uses of a waterbody, the numerical and narrative water-quality criteria necessary to protect the uses of that particular waterbody, and an antidegradation statement

water table the upper limit of the part of the soil or underlying rock material that is wholly saturated with water; in some places an upper or perched water table may be separated from a lower one by a restrictive layer; the level of groundwater

watershed the geographic area that drains to surface waterbodies; a watershed generally includes lakes, rivers, wetlands, estuaries, surrounding landscape, and contributing groundwater

watershed approach a watershed approach is a coordinating framework for environmental management that focuses public and private sector efforts to address the highest-priority problems within hydrologically defined geographic areas, taking into account groundwater and surface-water flow

watershed-based trading a market-driven approach that involves trading arrangements among point source dischargers, nonpoint sources, and indirect dischargers in which the buyers purchase pollutant reduction at a lower cost than what they would have to spend to achieve the reduction themselves; sellers provide pollutant reductions and may receive compensation; the total pollution reduction must be the same or greater than what would have been achieved if no trade occurred

watershed management plan a document by the watershed partnership that identifies all natural resources, the problems and concerns impacting those resources, management approaches, and strategies to address those problems, goals, and objectives for the proposed activities, and the schedules and milestones for the activities

watershed management process a four-step process, implemented through a partnership, that supports a flexible, dynamic approach for developing and implementing local solutions and watershed basis.

watershed monitoring monitoring primarily designed to sample and assess the characteristics and condition of a watershed, or to sample and assess specific entities on a watershed basis

watershed project a group of activities undertaken in a geographic area to restore the beneficial uses of a waterbody already affected, degraded, or threatened by point or nonpoint source pollution

watershed restoration action strategy a watershed restoration action strategy (WRAS) documents the most important causes of water pollution and resource degradation, details the actions that all parties need to take to solve those problems, and sets milestones by which to measure progress

weight of scientific evidence considerations in assessing the interpretation of published information of toxicity: quality of testing methods, size and power of study design, consistency of results across studies, and biological plausibility of exposure–response relationships and statistical associations

wetlands areas that are inundated or saturated by surface or groundwater at a frequency or duration sufficient to support, and under normal circumstances do support, a prevalence of vegetation typically adapted for life in saturated soil conditions

wind erosion the detachment and transportation of soil by wind

Index

A

Accountability, 152
Acid deposition, 34, 110
Action stage, *see* Implementation
Activities, Bennett's hierarchy of evidence, 151, 153, 154
Adaptability, management practice parameters, 138
Adaptive management, 155
    evaluation, 155
    goals/targets for, 97
    monitoring, 165
    restoration issues, 116
Administrative costs, 126
Administrative entities, 7
    management plan document contents, 123
    monitoring reports, 179–180
    partnerships, 10–11
    political, *see* Political boundaries/jurisdictions
Administrative goals, 155
Administrative performance evaluation, 152
Administrators, 57
Adoption process, social capacity building, 211
Aerial photography, 67, 82
Agencies and organizations
    implementation team, 129
    model process, 7
    partnership participation, 39
AGNPS, 186
Agreements, 56–57
Agriculture
    assessment phase, 70, 83–84
        data collection, 69–70
        designated use definition, 74
    CORE 4 system, 100–101
    GIS inventory, 199
    land use categories, 32, 33–34
    modeling, 187
        CREAMS, 188
        GIS, 197
    monitoring, 166–167
    parcel-level analysis, 135–137
    partnerships, 40
    planning process, 103–106
    pollutant sources, 2, 3, 30, 31, 77, 78
        nutrients, 22, 34
        pesticides, 32
        sediments, 3, 31
    Ythan project, 100
Algae blooms, 10
Alliances, *see* Partnerships
Alternative evaluation approaches, 156–157
Alternative management options
    modeling, 185
    planning phase, 95, 96
Alternatives analysis, planning, 118
Alternatives development, assessment phase, 14
Ambient monitoring, 169
Analysis
    assessment phase, 71–77
    failure of previous plans, 93
    models, 189, 195
    parcel-level, 135–137
Analytical method delineations, 65
Animal/livestock operations, 30, 77, 88
    agricultural land use, 83–84
    critical area identification, 84
    nutrient loading, 22, 34
    pollutant sources, 77
    three-tier analysis, 75
Annual review, 224
Approval, planning phase, 17
Aquatic life
    habitat changes and, 22
    Lower Big Rib River resource conditions, 88
    phosphorus loading and, 22
    physical characteristics and, 22
    sediment effects, 3
Aquifers, 70, 110
Arbitrary fixed-radius method, 64
ARC/INFO, 194
Area of plan, *see* Geographic area/scale
Assessment, 227
    monitoring and assessment issues, 163, 165, 169
    recommend process phases, 11
    of results/project performance, 18, 155; *see also* Evaluation
Assessment and problem identification phase, 13–15, 61–88, 155, 227
    analysis, 71–77
    rapid resource appraisal and visual inventories, 46

251

three-tier, 74–77
benchmarks for, 12
communications, 62–63
criteria for judging, 61
critical areas, 84–86
defining watershed, 65–67
implementation issues, 134–135
inventory stage, 68–71
management model, 5
management plan document contents, 124
model process, 6
monitoring and assessment issues, 165, 169
perceptions of problem and identification of concerns, 62
planning issues, 95, 118
pollutant source assessment, 77–84
    agriculture, 83–84
    urban, 80–83
project scope, 67–68
public outreach process, 209
scope/level of, 61–62
synthesis of information, 86–87, 88
Assessment of results, *see* Evaluation
Atmospheric deposition, 34–35
    nutrient sources, 22
    pollutant sources, 77, 78
    sources of pollution, 2
Attitude trend surveys, 70
Audience identification
    evaluation
        Bennett's hierarchy of evidence, 151
        targeting information for, 160
    social capacity building, 206
Authority for implementation
    planning issues, 107, 118
    reasons for failure of previous plans, 93, 94
Automobiles, 33, 34
AUTO-QI, 188
Awareness stage, public involvement, 222

B

Bacterial pollution, 2, 177
Bank features, *see* Stream banks
Barnegat Bay, New Jersey, 114–115
Barriers, 149–150
Base flow, 29
Baseline conditions
    model selection criteria, 191
    for monitoring, 166, 168, 169
Basins, 23
    assessment, scale of, 63
    drainage, 24
BASINS (program), 185, 191

BATHTUB, 190
Beaches, 2, 3
Behavior changes, 151, 152, 153, 154, 217
Bellingham, Washington, 133
Benchmarks, 12, 147; *see also* Goals/objectives/milestones/targets
Beneficial use impairment, 73–74
Bennett's hierarchy of evidence, 150–156
    components of, 150–153
    impact, 151–152
    implementation, 154–156
Best management practices, 211
Better Assessment Science Integrating Point and Nonpoint Sources (BASINS), 185, 191
Big Ditch Watershed, 34
Biological community
    assessment phase, 68, 73
        analysis, 72
        urban runoff pollution and, 81
    management plan document contents, 123
    monitoring, 167–168, 175
Bioretention cells, 138
Black Creek Project, Indiana, 84
Boardman River, Michigan, 66
Body contact, designated use definition, 74
Bottom-up planning process, 15
Boundaries, administrative/political, *see also* Political boundaries/jurisdictions
    hydrological versus political, 66
    management plan document contents, 123
Boundaries, watershed, *see also* Geographical area/scale
    hydrologic unit codes and, 23–24
    management plan document contents, 123
    management zone, 103–104
    project scope, 66, 68
Brainstorming, conflict resolution, 55
Brooks Creek, Michigan, 187
Budget/funding, 228; *see also* Costs
    committee structure, 43
    data collection constraints, 71
    evaluation phase, 18, 149, 152
    failure of previous plans, 93
    implementation, 17, 102, 130, 139
    institutional arrangements/organization, 40, 41
    local government participation, 59
    management plan document contents, 125–126, 127
    master schedule, 131
    models
        GIS, 197, 198
        selection of, 191
    monitoring program, 179

Index

plan elements, 119
plan implementation approaches, 102
RESULTREACH steps, 157
wetland and riparian zone conservation, 115
Buffer requirements, 110
Build-out analysis, 109–110, 112–113, 187
Bureau of Reclamation, 10
Business interests, partnerships, 40
Buttermilk Bay, Massachusetts, 110, 111, 112, 113

C

Calibration, model, 189, 190
Calculated fixed-radius method, 64–65
Canopy, 72
Capital costs, 126, 138
Cause-effect relationships
 evaluation process, 144
 monitoring data requirements, 171
Census data, 69–70
Chair, technical advisory committee, 47
Chambers of commerce studies, 70
Channel morphology, *see* Stream channel
Chemical, runoff, and erosion from agricultural management systems (CREAMS), 188
Chemical characteristics, 123
Chemical monitoring, 167–168
Chemical pollution, 2
Chemical variables, monitoring, 175
Chesapeake Bay, 34
Citizen advisory committees (CAC), 43, 47–48
Citizen interviews, 70
Citizen monitoring, 176–177
Citizens groups, monitoring goals, 164
Clean Lakes Program, 37
Clean Water Act, 2, 4
Climate, 66, 68, 174; *see also* Precipitation
Coalitions, *see* Partnerships
Coastal water quality monitoring program, 177
Cobbs Creek, Pennsylvania, 108–109
Cold water, designated use definition, 74
Collaboration
 partnership operations continuum, 52
 Western Governors' Association principles, 12
Combined sewer overflows, 78
Comment state, public involvement, 222
Committees, partnership structure, 42–49
Communications
 assessment phase, 62–63
 committee structure, 43
 conflict resolution, 55

evaluation process, 143–144, 150
 reports, 159–160
 target audience, 153
 failure, reasons for, 53–54
 management work plan for, 46
 monitoring program, 164, 179
 partnership operations, 50, 52
 planning, 120
 with public, 213–218
 social capacity building, 204, 205
 steering committee responsibilities, 44
Community, *see also* Public participation and support
 assessment, scale of, 63
 commitee structure, 43
 implementation, 132–135
 local government participation, 58
 partnerships, 38, 42–43
Community water supplies, 70
Compliance
 evaluation process, 146
 monitoring, 168, 170
Computer programs
 digital terrain model (DTM), 64
 modeling, *see* Models
Condition monitoring, 168, 169
Conestoga Headwaters Project, Pennsylvania, 173
Confidence, management practice parameters, 138
Confirmation stage, public participation, 213
Conflict resolution
 evaluation process, 143–144
 failure, reasons for, 53–56
 monitoring goals, 164
 partnership operations continuum, 52
Consensus building/decision making
 partnerships, 37, 40, 49–50
 steering committee responsibilities, 44
 Western Governors' Association principles, 12
Conservation Development Evaluation System (CeDES), 134
Conservation groups, 40
Conservation Technology Information Center (CTIC)
 Managing Conflict, a Guide for Watershed Partnership, 55
 partnership failure, reasons for, 52–53
 wetlands benefits, 113
Consistency, monitoring data requirements, 171
Constructed wetlands, 117
Coordination, partnership, 41, 51–52
Coordinator, paid, 41
CORE4 system, 100–101, 135, 228
Corpus Christi Bay, Texas, 186, 195, 197

Costs, *see also* Budget/funding
    assessment/data collection phase, 14, 63, 69, 71
    partnerships and, 38
    planning constraints, 96
    pollution control, 101
    restoration, 115
Costs and benefits/cost effectiveness
    evaluation criteria, 149
    phosphorus reduction, 99
    priority setting, 99
    urban area planning objectives, 107
    Western Governors' Association principles, 13
    management plan document contents, 127
Court Creek, Illinois, 79
Craddock Creek, 174
CREAMS (chemical, runoff, and erosion from agricultural management systems), 188
Criteria/standards, 1–2, 12
    assessment phase, 61
    evaluation process, 148, 158
    implementation plan, 132
    modeling data, 187
    monitoring program, 170, 173–174, 178
    national, 12
    objectives, *see* Goals/objectives/milestones/targets
    regulatory, single-purpose, 1–2
    restoration issues, 116
    water quality, 61, 123, 173
Critical areas
    assessment phase, 84–86, 87
    model applications, 186
    National Pollutant Discharge Elimination System (NPDES), 77
Critical loading rate, 110, 112
Critical site approach, 104, 105
Critical slopes, 110
Cropland, 34, 83–84
CTIC, *see* Conservation Technology Information Center
Cultural factors, 227
    failure of programs, reasons for, 54
    social capacity building, 211, 212
Cultural resources, data collection, 70
Cumulative impacts, 5

D

Darby watershed, 108–109
Data collection and analysis, 229
    adaptive management, 155
    assessment phase, 13–14, 62, 63
    database size and, 71
    frequency of, 69
    inventory, 68
    evaluation phase, 18, 149, 159, 160
    models, 187, 188
        GIS benefits, 195
        output analysis and interpretation, 189
        selection of, 191
    monitoring, 18, 163, 170, 171, 177–178
        baseline studies, 166
        quality of, 178–179
        RESULTREACH, 157
    Western Governors' Association principles, 12
Data gaps, 68, 100
Davis Creek, 66
Dead zone, Gulf of Mexico, 10
Decision making
    evaluation process, 150, 160
    failure of previous plans, 93
    management model, 5
    modeling and, 191–192, 195, 199
    partnerships, 49–50, 57
    planning, 95, 118
    public participation, 211, 223
    restoration issues, 116
    steering committee responsibilities, 44
Definitions, management plan document contents, 127
Definition of problem, monitoring and assessment issues, 165
Delaware River (New York) Model Implementation Project, 166
Delegated implementation, 132–133
Delegation of responsibility, 42
Demographic data, 68
Demonstration projects, 7, 147–148, 152, 228
Design
    assessment level for, 61–62
    evaluation plan development, 159
    model purpose, 185
    monitoring program, 178
    public participation, 223
    social capacity building, 206
Designated uses, 74; *see also* Land use
Developers, partnerships, 40
Development, *see* Economic growth and development
Diagnostic studies, data collection, 70
Digital terrain model (DTM), 64
Dimensions of model, 190
Division of labor, partnership structure, 42
Document, management plan, 123–127

Index

Documentation
  evaluation process, 142–143, 144, 151, 152, 157
  implementation requirements, 132
  management plan document contents, 127
  modeling, 193
  model selection, 191
  monitoring and assessment issues, 165
  planning, 95
  public participation, 223–224
  reasons for failure of previous plans, 93
  RESULTREACH, 157
  results, 142–143
Document repository, 49
Double Pipe Creek RCWP, 12, 130, 135
Drainage area, 24–25, 26
  analysis of individual parts, 73
  effective, 28
  integration of factors for planning, 86
  land use and pollutant types, 79
  management plan document funding, 126
  partnership requirements, 10–11
  system overview and linkages, 64
  and watershed size, 66
Drainage basins, 24
Drainage density, 25
Drawdown, 30
Drinking water, 2
  defining watershed, 64–65
  monitoring, 174
Durability
  evaluation criteria, 148
  management practice parameters, 138
Duration
  flow, 29
  IDF curves, 28
  monitoring, 170, 178
  precipitation, 27
  sampling
Dynamic equilibrium, 31, 32

E

Ecological impacts, *see* Environmental impacts; Environmental indicators
Ecological integrity assessment, 72
Economic growth and development
  build-out analysis, 112, 113
  coastal, with resource conservation, 114–115
  implementation, 134
  data collection, 70
  partnerships, 40
  planning phase, 15–16, 96
Economic resources, 118

Economics, 227, 228
  assessment data, 63, 68, 69
    analysis of individual parts, 73
    integration of factors for planning, 86
    system overview and linkages, 64
  evaluation process, 143–144, 150
  planning issues, 96, 118
  Western Governors' Association principles, 13
Edge-of-field monitoring, 169, 175
Education
  evaluation process, 143–144
  partnership operations, 50
  plan implementation approaches, 102
  public participation, 204, 209–213
    evaluation (confirmation) stage, 213
    implementation stage, 213
    knowledge stage, 211–212
    persuasion stage, 212
  social capacity building, 205
Education symposium, 219
EFDC model, 190
Effective drainage area, 28
Effectiveness, *see also* Evaluation; Performance/results
Effective watershed, 28
Elected officials, partnerships, 40
Elevation, 25
Endangered species, 70
Energy
  ecological integrity of streams, 72
  factors affecting quality and function of resources, 21, 22
  urban runoff and, 79
  watershed, 22
Enrichment ratio, 86
Environmental groups, partnerships, 40
Environmental impacts, 46
  assessment data, 63–64
  monitoring program, 178
Environmental indicators
  data collection, 70
  goals/targets/benchmarks, 97
  three-tier analysis, 76
Environmental Systems Research Institute, 194
Ephemeral flow, 29
Equilibrium, dynamic, 31–32
Erosion, *see also* Soil erosion/loss; Stream banks, erosion
  critical area identification, 84
  geomorphologic processes, 30–31
  Lower Big Rib River resource conditions, 88
  three-tier analysis, 75
Estuaries
  drainage networks, 24–25, 26

local government participation, 58
model dimensions, 190
pollutants, 3
sources of pollution, 2
Ethan Creek, 173–174
Evaluation, 16–17, 141–161, 227
  adaptive management, 155
  alternative approaches, 156–157
  assessment levels and, 62
  barriers, 149–150
  committee structure, 43
  implementation phase, 17
  information collection, 141–142
  levels of, Bennett's hierarchy of evidence, 150–156
    components, 150–153
    evidence of impact, 151–152
    implementation, 154–156
  management model, 5
  management plan document contents, 125
  milestones, 141
  models, 7, 187; *see also* Models
  monitoring, 171, 172
  planning phase, 17, 119, 158–159
  public participation, 213
  recommend process phases, 11
  reporting, 159–160
  types of, 144-149
    formative evaluation, 144
    impact, 148–149
    outcome, 146–148
    process evaluation during, 145–146
Evaluation team
  GIS use, 199
  operations committee, 48
  watershed phases, 155
Evaporation, 27
Event mean concentration (EMC), 186
Everglades, 2
Existing programs and projects
  assessment phase, 64–65
  data collection, 70
  reasons for failure of previous plans, 93

F

Failure of programs, reasons for, 52–53, 92–94
Fall Creek, Indiana, 23
Feasibility
  data collection, 70
  evaluation standards, 158
Federal agencies, 227, *see also* Regulatory agencies
  coordination with, 44

partnerships, 57
technical advisory – composition, 46–47
Feedback
  evaluation phase, 17–18
  public outreach process, 220
Feedback dynamics, implementation, 139–140
Fertilizers, 30; *see also* Nutrients
  build-out analysis, 110, 112
  nutrient sources, 22
  Ythan River project, 99–100
Field scale
  implementation, 135–137
  monitoring, 175
Field trips
  assessment phase, 86
  public education, 211–212
Financial commitments, *see* Budget/funding
Financial institutions, partnerships, 40
Financial support, *see* Budget/funding
First-come, first-served approach, 103–104
First-order streams, 24, 26
Fish and wildlife, 2
  assessment phase, 70, 72
  cumulative impact of pollutants, 3
  designated use definition, 74
  Lower Big Rib River resource conditions, 88
  monitoring, 174
  pesticides and, 32
  three-tier analysis, 75
  urban area planning, 108
  wetland-to-watershed ratio and, 113
FISRWG, 73
Fixed-radius methods, 64–65
Flood control, 126
Flooding
  Lower Big Rib River resource conditions, 88
  map information, 67
  wetlands and, 113, 114
  wetland-to-watershed ratio and, 113
Floodplain, restoration and, 116–117
Florida Everglades, 2
Flow, *see* Stream flow
Fluvial geomorphology, 32
Focus groups, 69, 215
Follow-up, implementation team, 137
Food web, 22
Forestry, 77
Forest Service, 10
Formal agreements, partnerships, 56–57
Formartine Partnership, 100
Format, management plan, 126, 127
Formative evaluation, 144
Forming process, 50

Index

Fragmentation of mission, regulatory agencies, 9–11
Friends of the Chicago River, 7
Funding, *see* Budget/funding
Fundraising, committee structure, 43

G

Gantt chart, 124
Garvin Brook, Minnesota, 84, 173, 199
Generalized watershed loading functions (GWLF), 186, 187, 191, 192
Geographic area/scale, 23, 24
    assessment data, 61–62, 63, 87
        defining watershed, 65
        project scope, 67–68
    implementation, 132–137
        community, 132–135
        field scale, 135–137
    management plan document contents, 127
    modeling, 186, 189
        GIS, 199
        selection of model, 191, 192
    monitoring, 166, 175–176, 179
        field, 175
        watershed, 175–176
    and partnership structure, 42
    planning, 94
    reasons for failure of previous plans, 92
    size of watersheds, 10–11
    transboundary issues, 229
Geographic areas, Western Governors' Association principles, 13
Geographic information systems (GIS), 4, 185, 188, 193–199
Geographic priorities, restoration, 116
Geographic resources analysis support system (GRASS), 195
Geomorphology, 30–31, 32
    analysis of individual parts, 73
    integration of factors for planning, 86
    mapping, 67
    and percolation, 30
    system overview and linkages, 64
    and watershed size, 66
GIS (geographic information systems), 4, 185, 188, 193–199
GLOBE program, 177, 219
Goals/objectives/milestones/targets, 228
    assessment phase, 12, 14
    conflict resolution, 55
    evaluation of results, 145–146, 148; *see also* Evaluation
    data, 152
    criteria, 148
    levels of, 153
    modification/revision of, 155–156
    setting, 155
    federal agencies, 10–11
    institutional arrangements/organization, 40
    management plan document contents, 124, 127
    master schedule, 131
    modeling, 185
    monitoring, 163–164, 173–174
    open endedness of, 46
    planning phase, 91, 96–97
        plan elements, 119
        priority setting, 99–100
    public participation, 206–207, 223
    rapid resource appraisal and, 45–46
    regulatory, failure to meet, 2
    standards, *see* Criteria/standards
Government
    model process, 7
    partnerships, 37, 40
    planning, 95
Government jurisdictions, *see* Political boundaries/jurisdictions
Government protected areas, map information, 67
Government regulation, *see* Regulation; Regulatory agencies
Grab sampling, 177
GRASS, 194
Great Lakes, 197
Great Swamp Watershed Project, 44
Green space, 190
Groundwater, 2
    analysis of individual parts, 73
    assessment data, 71
    hydrologic cycle, 29–30
    integration of factors for planning, 86
    pesticides, 32
    surface water and, 24
    system overview and linkages, 64
    water cycle, 26, 27
Groundwater drainage basin, 24
Groundwater recharge
    modeling, 190
    wetland-to-watershed ratio and, 113
Growth and development, *see* Economic growth and development
Guidance for Specifying Management Measures for Nonpoint Sources in Coastal Waters, 138
Gulf of Mexico, 10
GWLF, 186, 187, 191, 192

H

Habitat, 3
  assessment, 72, 73
  factors affecting, 21, 22, 79
  Lower Big Rib River resource conditions, 88
  restoration, 116
  urban runoff and, 79
  wetland-to-watershed ratio and, 113, 114
Haines Creek Project, 173
Hardware, model selection, 191
Hawthorne effect, 153
Hidden resource, groundwater as, 30
Hierarchy, *see* Priorities
High-flow regimes, 22
Highland Silver Lake, Illinois, 86, 143, 173
Historical data, 69
Homeowners, *see* Land ownership
Hortonian runoff, 28
Housing data, 69
HSPF, 187–188, 190, 191
HSPF/BASINS, 191
Hudson River Basin, 24
Human resources, planning, 118
Huron River Watershed Council, 216
Hydraulic models, 184–185
Hydrocarbons, 33
Hydrologic codes, watershed units, 23–24
Hydrologic cycle, 25–30
  groundwater, 29–30
  precipitation, 26–28
  runoff, 28–29
  stream flow, 29
Hydrology
  assessment data, 70–71
    analysis of individual parts, 73
    synthesis of interacting factors, 86
  data collection, 70
  integration of factors for planning, 86
  management plan document contents, 123, 127
  model calibration/validation, 190
  model selection, 190, 192
  system overview and linkages, 64
  three-tier analysis, 75
  urban area planning, 107
  watershed strategy development, 100
  watershed-to-wetland ratio and, 22
Hydromodification, 2
Hypoxia, 10

I

Idaho RCWP project, 33
Identify and reply communications strategy, 215
IDF (intensity, duration, frequency) curves, 28
Illinois EPA, watershed scale recommendations, 66
Illinois RCWP project, 33
Illinois River, 23
Illinois State Water Survey, 22
Impervious cover
  build-out analysis, 109–110
  implementation phase, 134
  management plan document contents, 123
  modeling, 190
  planning objectives, 107
  stream quality categories, 109
  urban runoff, 80–83
Implementation, 16, 129–140, 229
  assessment benchmarks, 12
  assessment levels and, 62
  committee structure, 43
  dynamics, 139–140
  funding, 139
  goals/targets/benchmarks, 97–98
  hierarchy/scale of, 132–137
    community, 132–135
    field scale, 135–137
  institutional aspects, 130, 132
  local government participation, 58
  management model, 5
  management plan document contents, 124, 126, 127
  management practice aspects, 138–139
  master schedule, 130, 131
  model process, 6
  monitoring, 165, 169, 170
  partnership operations, 50
  planning process, 102–115, 119
    agriculture, 103–106
    modeling data, 187
    urban, 107–113
    urban riparian and wetland management, 113–115
  priority setting, 98–100
  public participation, 213, 222–223
  recommend process phases, 11
  results, 142–143, 145–146; *see also* Evaluation
  social capacity building, 207
  volunteer contributions, 139
Implementation team
  evaluation data, 152
  GIS use, 199
  operations committee, 48
  public education, 212
Incinerators, 34
Indian River, 78

Index

Indices, surrogate, 86
Industrial use, 74, 77
Infiltration, 2, 25, 27, 30
Information, *see also* Data collection and analysis
    assessment phase, 86–87, 88
    plan implementation approaches, 102
    public, 203–204
Information technology, 4
Infrastructure, map information, 66
Injection of wastewater into groundwater, 30
Inputs
    Bennett's hierarchy of evidence, 151
    evaluation data, 153, 154
    model requirements, 188–189
Institutional arrangements, organization, 40
Institutional aspects
    implementation, 130, 132
    partnerships, 56–57
Institutional changes, evaluation process, 143–144
Institutional characteristics of watershed organization, 40–41
Interconnectedness, 5
Interdisciplinary technical advisory committee, 46–47
Interest groups, *see* Stakeholders
Interlocal agreements, 44
International Joint Commission for Great Lakes, 197
Interviews, 69, 70, 144, 215
Inventories, 13–14, 62, 64
    planning, 118
    visual, 46
Inventory stage, 68–71
Inventory team, 62
Inverted factor analysis, 58
Investigations, model process, 7
Irrigation, 33, 79–80

J

Johnson-Sauk Trail Lake, Illinois, 117, 166
Jurisdictional boundaries, *see* Political boundaries/jurisdictions

K

Kalamazoo River, 66
Knowledge, attitude, skills, and ability (KASA), 150, 151, 152, 153–154
Knowledge stage, public participation, 211–212

L

Lake Champlain National Nonpoint Source Monitoring Project, 199
Lake Emily, 174
Lake Le-Aqua-Na, Illinois, 117, 149
Lake McCarrons, Minnesota, 118–119
Lakes
    drainage networks, 24–25, 26
    model dimensions, 190
    natric soil loading, 33–34
    pollutants, 2, 3
    sources of pollution, 2
Lake Superior, 34
Lake Wobegon factor, 93–94
Lamington River, 24
Land area, *see* Geographic area/scale
Land conversion, three-tier analysis, 75
Land management, *see* Management practices
Land ownership
    data collection, 70
    GIS inventory, 199
    implementation phase, 17
    parcel-level analysis, 135–137
    partnerships, 40
    planning process, 94
    social capacity building, 211
    wetland and riparian zone conservation issues, 114
Land parcel, implementation scale, 135–137
Landscape approach, 105–106
Landscape function, hydrologic unit codes and, 23–24
Landscape-level monitoring, 175
Land treatment, monitoring, 175, 176
Land use, 32–35
    adopted management, 101
    assessment data, 68
        collection, 70
        data types, 63
        synthesis of interacting factors, 86
    and bank erosion, 79–80
    build-out analysis, 112, 113
    designated use definition, 74
    ecological integrity of streams, 72
    GIS inventory, 199
    impervious cover estimation, 82
    implementation, 133–134
    management aspects, 31
    management plan document contents, 123
    map information, 66, 67
    models
        calibration/validation, 190
        GIS, 197
        selection of, 191, 192

utilization of, 186
monitoring, 174, 175
planning
    local government participation, 58
    urban, 110
planning constraints, 96
and pollutant types and concentrations, 77, 78, 79
public awareness, 228
and water quality, 21–22
watershed processes, 32
    agriculture, 33–34
    urban, 33
and watershed size, 66
wetland and riparian zone conservation issues, 114
Land use regulations
    build-out analysis, 110, 112
    reasons for failure of previous plans, 92
LaPlatte River Basin, Vermont, 99
Largo, Maryland, 138
Lawsuits, 40
Lead agency/organization, 44–45, 56
Leadership, 228
    committee structure, 43–44
    evaluation issues, 150
    institutional arrangements/organization, 40
Le-Aqua-Na Lake, Illinois, 117, 149
Learning, RESULTREACH, 157
Legislation
    data collection, 70
    failure to meet water quality goals, 2
    historical, 4
    wetland and riparian zone conservation, 114–115
Library, assessment data, 62–63
Little Conestoga Watershed Alliance, 38–39
Livestock, *see* Animal/livestock operations
Loading/loading rate, *see also* Pollutants, loads and sources
    assessment phase, 79
    build-out analysis, 110, 112
    GWLF, 186, 187, 191, 192
    modeling, 187, 191
    monitoring, 167
Load reduction estimation, 78
Local authorities, political involvement, 57–58
Local initiatives, 40–41
Local ownership, *see also* Land ownership
    reasons for failure of previous plans, 92
    scoping process, 62
Local partnerships, 10, 40–41, 57–58
Local programs, reasons for failure of previous plans, 93
Local rainfall curves, 28

Local regulations, 70
Long-term impact monitoring, 169
Lost River, West Virginia, 185
Lower Big Rib River, 88, 97
Lower Fox River, Illinois, 23
Lower Hudson River Basin, 24
Low-flow regimes, 22

M

Macatawa Watershed Project, 192
Madison River, Montana, 79
Management, 4–8, 31–32
    decision making authority of managers, 57
    federal regulations, history of, 4–5
    goals, *see* Goals/objectives/milestones/targets
Management activities
    assessment data, 63
    costs per activity, 101
    evaluation process, 152, 159
    GIS use, 199
    model calibration/validation, 190
    monitoring, 169, 175
Management-level approach, models for, 186
Management plan document, 123–127
    elements of, 123–126
    format/template, 126, 127
    funding, 126
Management planning, *see* Planning
Management practices
    evaluation process, 145–146
    implementation, 138–139
    guidance manuals, 140
    management plan document contents, 124, 125
    monitoring data, 174
Management process, 9–19
    assessment phase, 13–15, 79, 86–87
    evaluation, 16–17
        data, 152
        modified Bennett's hierarchy, 154
    implementation, 16
    limits and scope of, 9
    phases of, 13–18
    planning, 15–16
    public participation incorporation into, 223–224
    regulatory agencies, fragmentation of mission, 9–11
Management systems, site-specific implementation, 137
Management zone basis
    integrated pollutant source and transport approach, 106

Index

proximity to water body, 105
watershed boundary, 103–104
Managing Conflict, a Guide for Watershed Partnership, 55
Managing Lakes and Reservoirs, 215
Mandated implementation, 132–133
Manure, 30, 83–84
Maps
   assessment phase, 66–67
   data collection, 70
   GIS uses, 195
   hydrologic unit codes, 23–24
   topographic, 64
Market research, 215–217
Markets, 13
Marshall Drain, 137
Master monitoring sites, 169
Master schedule, 130, 131
Mathematical models, 184–185
Maumee River, Ohio, 147–148, 174
McCarrons Lake, Minnesota, 118–119
Measures/benchmarks/criteria/standards, *see* Criteria/standards
Media, public outreach, 206, 207, 219, 220, 221
Merrimack River Watershed, Massachusetts, 194
Metabolism, 22
Metals, 3, 34
Methodology, evaluation, 150
Microwatershed, 23
Middleton, Wisconsin, 190
Milestones, 12; *see also* Goals/objectives/milestones/targets
Minimum detection change (MDC) level, 176
Minimum standards, regulatory approach by, 2
Minnesota RCWP Project, 34
Mission fragmentation, regulatory agencies, 9–11
Mission statement
   and data collection, 69
   partnership, 45
   steering committee responsibilities, 44
Mississippi River, 23
   nonpoint source runoff affecting Gulf of Mexico, 10
   stream order, 25
Mitigation corridors, 110
Mixed use systems, 32, 191
Models, 183–199
   application, 186–190
   category, 184–185
   documentation plan, 193
   geographic information systems (GIS), 193–199
   monitoring program data needs, 178
   selection of, 190–192

Modified universal soil loss equation (MUSLE), 197
Monitoring, 163–180, 227, 228
   committee structure, 43
   development of effort, 177–179
   evaluation phase, 18, 149, 152, 154
   frequency, duration, data analysis, intensity, and scale, 170
   GIS use, 199
   goals/targets/benchmarks, 97
   modeling and, 187, 192
   operations committee, 48
   plan elements, 119, 125
   purposes, 168–175
      compliance, implementation, and regulation, 170
      evaluation/effectiveness, 171, 172
      monitoring by objectives approach, 173–174
      problem investigation, 170
      water quality, processes in, 171, 173
   reporting, 179–180
   scale of, 175–176
      field, 175
      watershed, 175–176
   types of, 167–168
   volunteer, 176–177
   watershed, 164–167
Monitoring by objectives approach, 173–174
Monitoring teams, 48, 199
Municipalities
   interlocal agreements, 44
   pollutant sources, 2, 3, 77–78
Municipal waste, 88, 100

N

Nansemond-Chuckatuck Project, Virginia, 171
National Pollutant Discharge Elimination System (NPDES), 77, 108
National Water Quality Inventory, 2
National Weather Service rain zones, 27
Natric soils, 33–34
Natural features, map information, 66, 67
Natural resources
   data collection, 70
   three-tier analysis, 74
Navigation, 74
Needs assessment, social capacity building, 209
Neighborhood solutions, 12
Nesting, watershed, 23, 24
Newspapers
   as data source, 69
   public relations, 206, 207, 219, 220, 221

New York City Staten Island Blue Belt project, 109
Nitrogen
  build-out analysis, 110, 111, 112, 113
  Gulf of Mexico, hypoxia in, 10
  sources, 34
  three-tier analysis, 76
  Ythan River project, 99–100
Noncompliance, evaluation process, 146
Nonpoint source pollution
  agriculture, 33
  atmospheric deposition, 34
  cumulative effects, 3
  evaluation types, 154
  ineffectiveness of regulation, 5
  modeling, 187
    BASINS, 185
    GIS estimation, 195, 197
  and Gulf of Mexico hypoxia, 10
  Lower Big Rib River, 88
  monitoring, 169
  nutrients as, 22
  precipitation, 28
  sources of, 77, 78
  surface runoff and, 29
Norming, 51
North Fork Embarrass River Project, Illinois, 141
Nutrients
  agricultural sources, 34
  build-out analysis, 110
  critical area identification, 84
  ecological integrity of streams, 72
  Illinois RCWP project, 33
  leading pollutants, 3
  Lower Big Rib River resource conditions, 88
  management practice parameters, 138
  monitoring, 166–167
  planning phase priority setting, 98
  pollutant sources, 78
  sources of, 30
  total maximum daily load, 187–188
  three-tier analysis, 75, 76
  and water quality, 21–22
  Ythan River project, 99–100

O

Objectives, *see* Goals/objectives/milestones/targets
Objectives, monitoring by, 173–174
On-site sewage disposal systems (OSDS), 78
Open spaces, 107
Operations, 137
  management plan document contents, 125
  management practice parameters, 138
  model purpose, 185
  partnerships, 49–56
  planning phase, 16
  steering committee responsibilities, 44
Operations committee, 43, 48–49, 129
Organic Administrative Act of 1897, 4
Organic chemicals
  atmospheric deposition, 34
  leading pollutants, 3
  threats to quality and sources of, 30
Organic matter
  ecological integrity of streams, 72
  and water quality, 21–22
Organization/agency, 228
  evaluation issues, 150
  model process, evaluation, 7
  partnerships, 39, 40, 41–56
    operations, 49–56
    structure, 42–49
    steering committee responsibilities, 44
    technical advisory committee composition, 46–47
Otter Lake, Illinois, 94
Outcome evaluation, 146–148, 158; *see also* Evaluation
Output variables, model selection, 191
Outreach, *see also* Public participation and support
  committee structure, 43
  implementation teams, 130
  management plan document contents, 125
  public participation, 205–208
Outreach team, 48
Owasco Lake, New York, 199
Ownership, *see also* Land ownership
  plan implementation approaches, 102
  planning constraints, 96
Oxygen levels, 21–22, 72, 88

P

Paperwork, *see* Documentation
Parcel-level analysis, 63, 135–137
Parking lots, 138–139
Parks, 108–109
Partial body contact, designated use definition, 74
Participants
  evaluation, 148
  public, *see* Public participation and support
Partnerships, 5, 37–59, 227–228
  adaptive management, 155
  building, 38–41
  definition, 38

Index

evaluation, 7, 148, 149, 153, 160
expectations of results, 145–146
four-stage development process, 50–51
implementation, 130
institutional aspects, 56–57
management plan document contents, 123, 126
model process, 7
monitoring goals, 164
monitoring reports, 179–180
organization, 41–56
    operations, 49–56
    structure, 42–49
planning, 95
    failure of previous programs, 92
    priority setting, 98–100
political involvement, 57–59
public involvement issues, 62–63
results, documentation of, 142
social capacity building, 209
Pathogens, 3, 21–22, 30, 78
Peak flows, implementation, 134
Pebbles Lake, 174
P8, 191
Perceptions of problem, 62
Percolation, 26, 30
Perennial streams, 29
Performance/results; *see also* Evaluation
evaluation, 148
management system, 137, 138–139
monitoring, 168, 170
    data for, 171
    selected characteristics of, 172
partnerships, 51
Permits
management plan document contents, 124, 127
violations, evaluation process, 146
Personality conflicts, 53
Persuasion stage, public participation, 212
Pesticides, 32, 78
Pheasant Branch, Wisconsin, 190
Philadelphia, Pennsylvania, 108–109
Phillips' hierarchy, planning strategy, 100–101
Phosphorus
agricultural sources, 34
cost effectiveness of reduction, 99
management practice parameters, 138
monitoring, 166–167
three-tier analysis, 76
and water quality, 22
Photosynthesis, 22
Physical features, 22
assessment phase, 68, 86
map information, 67

Physical models, 184–185
Physical monitoring, 167–168
Physical processes, *see also* Hydrology
modeling, 187–188
monitoring, 175
Pilot programs, evaluation data, 152
Pittsfield City Lake, Illinois, 23
Plan document, 123–127
Planning
assessment, scale of, 63
data for, 70
evaluation, 158–159
implementation, 134–135
institutional arrangements/organization, 40
integration of elements for, 86
local government participation, 58
management model, 5, 6
management plan document contents, 127
modeling and, 185, 192
monitoring priorities, 164
partnerships for, 37, 45, 50
public participation, 206, 223–224
recommend process phases, 11
steering committee responsibilities, 44
targets, load reduction, 78
Planning committee, 43, 46
assessment of watershed, 62
data collection/inventory, 69
evaluation
    data for, 152
    levels of, 150
    modification/revision of goals and objectives, 155–156
technical advisory committee and, 47
Planning phase, 15–16
modeling data, 187
monitoring and assessment issues, 165
Planning process/plan development, 91–120
background, 92–94
failure, reasons for, 92–94
goal definition, 91
implementation approaches, 102–115
    agriculture, 103–106
    urban, 107–113
    urban riparian and wetland management, 113–115
progression of, 94–101
    planning phase, 96–98
    priority setting, 98–100
    watershed strategy development, 100–101
restoration, 115–117
results, 117–120
    parts of plan, 119–120
    schedule/timeframe, 117–118
Plant life, 3; *see also* Vegetation

Point sources
  evaluation process, 146
  injection of wastewater into groundwater, 30
  management plan document contents, 123
  modeling, BASINS, 185
  permits, management plan document contents, 127
  sources of, 77, 78
Political boundaries/jurisdictions
  coordination with state, local, tribal governments, 44
  hydrological boundaries and, 66
  map information, 66
  management plan document contents, 123
  Western Governors' Association principles, 13
Politics
  bureaucratic, 53
  evaluation issues, 150
  failure of previous plans, 92, 93, 94
  partnerships, 57–59
  rapid resource appraisal and, 46
Pollutants, load and sources, *see also* Loading/loading rate
  assessment data, 71, 72, 73, 77–84
    agriculture, 83–84
    critical area identification, 84
    location of, 87
    types of, 63
    urban, 80–83
  effective drainage area load and, 28–29
  management plan document contents, 124
  modeling, 185, 187
    BASINS, 185
    selection of model, 191
  monitoring parameters, 166, 177–178
  planning phase
    priority setting, 98
    watershed strategy development, 100–101
  present threats, 2
  stream flow for estimation of, 29
  three-tier analysis, 74
  urban runoff and, 81
  and water quality, 21–22
  wetlands and, 114
Pollutant source and transport approach, management zone, 106
Population, 69, 83, 96
Potential pollutant index, 86
Power struggles, 53
Precipitation
  assessment data, 71
  atmospheric deposition, 34
  build-out analysis, 110, 112
  data collection, 70
  drainage density, 25
  ecological integrity of streams, 72
  factors affecting watershed health, 22
  hydrologic cycle, 26–28
  management considerations, 31
  model selection, 192
  monitoring parameters, 177
  and stream flow, 29
  water cycle, 25
  water quality changes, 174
Predicative models, 110
Predictive tools, *see* Models
Press coverage, 206
Priorities
  implementation, need for demonstrable results, 130
  planning, 98–101, 118
  public outreach, 207
  restoration, 116
Priority resources, 17
Privacy issues, 18
Problem definition, *see also* Assessment and problem identification phase
  failure, reasons for, 53–54
  management plan document contents, 124
  monitoring and assessment, 165, 169
  planning, 118
  phases of adaptive management, 155
Problem investigation monitoring, 168, 170
Problem solving
  evaluation types, 154, 155
  local government participation, 58
Procedural decision making, 44
Processes, watershed, *see* Watershed processes
Process phase, evaluation, 145–146, 154
Product of restoration effort, evaluation, 148
Program evaluation, *see* Evaluation
Program impact monitoring, 169
Program milestones, evaluation process, 145–146
Progress monitoring, 152, 169; *see also* Evaluation
Project area, *see* Geographic area/scale
Property damage, 2
Propriety, evaluation standards, 158
Protection, costs per activity, 101
Proximity to water body, management zone, 105
Public interest groups, planning phase, 17
Public opinion surveys, data collection, 70
Public participation and support, 7, 203–224, 228
  annual review, 224
  assessment data availability, 62–63, 79
  communications, 213–218
  education, 79, 209–213
    evaluation (confirmation) stage, 213
    implementation stage, 213

Index

knowledge stage, 211–212
persuasion stage, 212
evaluation process, 143–144
evaluation problems, 149
funding levels and, 142
general principles of, 218–223
GIS benefits, 195
implementation phase, 120, 130, 131
management plan document contents, 125
monitoring goals, 164
outreach, 205–208
partnership organizational structure, 45
planning for successful participation, 102, 120, 223–224
public and stakeholders, 205
social capacity building, 204
Public service announcements, 219
Public water supply, designated use definition, 74
Puget Sound, Washington, 133

Q

QUAL2E, 190
Quality assurance/quality control, monitoring program data, 178–179
Quality of resource, planning phase, 15
Questionnaires, 215

R

Racine, Wisconsin, 83
Rainfall cycles, *see* Precipitation
Rain gardens, 138
Rain zones, 27
Random sampling, stratified, 168
Ranking, priority setting, 98–100
Rapid resource appraisal (RRA), 45–46
Raritan River Basin, 24
Raster-based GIS, 195, 196
Rawls Creek, South Carolina, 213
Reactions, Bennett's hierarchy of evidence, 151, 153, 154
Receiving water flows, seasonal factors, 70
Recharge areas, defining watershed, 64–65
Recommendations, reasons for failure of previous plans, 93
Recommend process, phases of, 11–12
Recreation, 74, 113
Region, 23
Regulation
  failure of previous plans, 93
  implementation approaches, 102
  monitoring, 170, 173

piecemeal, 1–2
partnership failure, reasons for, 53
planning constraints, 96
Regulatory agencies, 227
  fragmentation of mission, 9–11
  partnerships, 37, 40
    decision-making authority, 57
    reasons for failure, 53
Reliability of data, 69
Remote imaging, 4, 67
Reports, *see also* Documentation
  assessment, 62–63, 70, 73
  evaluation, 150, 159–160
    criteria, 148
  plan development, 159
  monitoring, 179–180
Repository
  assessment data, 62–63
  document, 49
Resource appraisal
  assessment phase, 45–46, 62–63, 64, 70
  planning, 118
Responsibility
  master schedule, 131
  partnership operations, 50
Responsiveness, monitoring data requirements, 171
Restoration
  costs per activity, 101
  implementation phase, 17
  planning phase, 15, 115–117
  urban area planning, 108–109
RESULTREACH, 157
Results, 117–120; *see also* Evaluation
  Bennett's hierarchy of evidence, 151, 152, 154
  implementation priorities, 130
  management practice parameters, 138
  Western Governors' Association principles, 12
Return on investment, RESULTREACH steps, 157
Rewards, Western Governors' Association principles, 12
Riparian zones
  build-out analysis, 110
  landscape approach, 104, 106
  planning process, 113–115
    site-specific, 136–137
    urban area, 108
  restoration, 116–117
Rivers and Harbors Act of 1899, 4
Rivers and streams, *see also* Riparian zones; Stream banks
  estimated number and length, 25
  groundwater and, 24

pollutants, 3
sources of pollution, 2
Roads
  impervious cover estimation, 83
  sediment sources, 31
Rockaway Creek, 24
Rock Creek, Idaho, 33, 79, 173, 175
Roles and responsibilities, management plan document contents, 125
Rules, partnership operations, 49–50
Rules of thumb, model calibration/validation, 190
Runoff, 2
  build-out analysis, 110, 112
  ecological integrity of streams, 72
  effective drainage area load and, 28–29
  factors affecting watershed health, 22
  and flow regimes, 79
  hydrologic cycle, 28–29
  management practice parameters, 138
  model selection, 192
  pollutant sources, 78
  precipitation analysis, 27
  regulation, problems created by, 2
  rural, 31
  snow-melt, 31
  tillage systems and, 136
  urban, 31, 81
    impervious cover and, see Impervious cover
    parking lot, 138–139
  water cycle, 26, 27
  and water quality, 21–22
Rural areas
  implementation scale, 135
  model selection criteria, 191
  runoff volume, 31
  sources of pollution, 30

S

Saginaw Bay, Michigan, 197
Salinity, monitoring, 177
Sampling (monitoring), 177
  goals of program and, 170
  monitoring program, 177
  interval/frequency of, 167–168
  site selection, 179
Sanitary sewers, pollutant sources, 78
Satellite imagery, 67
Scale of monitoring, 170, 179
Scale of watershed, 23, 24; see also Geographic area/scale
Schedule, see Timeframe

Science-based approach, Western Governors' Association principles, 12
Scientific limitations, planning constraints, 96
Scope of assessment, 61–62, 67–68; see also Geographic area/scale
Scoping process, 62; see also Assessment and problem identification phase
Screening-level models, 86, 185
Seasonal cycles
  ecological integrity of streams, 72
  model selection, 192
Seasonal data, 70
Second order streams, 25, 26
Sediment/sediment load and transport, 184
  dynamic equilibrium conditions, 32
  geomorphologic processes, 30
  irrigation and, 79–80
  Lower Big Rib River resource conditions, 88
  management practice parameters, 138
  models
    calibration/validation, 190
    dimensions of, 190
    GIS, 197, 198
  monitoring, 166–167
  pollutant sources, 78
  sources of, 3, 33
  stored nutrients, 22
  three-tier analysis, 76
  and water quality, 21–22
Seepage, geology and, 30
Septic systems, 30, 75, 217–218
Service learning, 219
Sewage treatment facilities, 78
Short-term monitoring, 169
Side effects, management practice parameters, 138–139
Signs of success monitoring, 154, 169, 171
Siltation, 3, 33
Silver Lake, Illinois, 33
Sinkholes, 34
Site-specific data, scale of assessment, 63
Size of watershed, 10; see also Geographic area/scale
Skokie River restoration, 117
SLAM, 191
Sludge, 30
Small Watershed Program, 4, 37
Snow melt runoff, 31
Social and Environmental Research Institute, 58
Social capacity building; see also Public participation and support
  partnership operations, 50
  wetland and riparian zone conservation, 114–115
Social capital, evaluation process, 143–144

Index

Social equity, 13
Social factors
   assessment phase, 68, 73
   integration of factors for planning, 86
   rapid resource appraisal and, 46
   system overview and linkages, 64
Social surveys, 215–216
Social values, *see* Values
Socioeconomic data, 69
Soil erosion/loss
   agricultural land use, 84
   geomorphologic processes, 30–31
   Lower Big Rib River resource conditions, 88
   monitoring, 174
   tillage systems and, 136
Soil moisture, water cycle, 27
Soil productivity measures, 61
Soil reserves, nutrient sources, 22
Soils
   data collection, 70
   GIS inventory, 199
   modeling data, 189
Soil saturation, runoff, 28
Sole source aquifers, build-out analysis, 110
Solids, suspended, 31, 76, 184
Source analysis, pollutants, 78
Source assessment, 77–84
Source identification, *see* Pollutants, load and sources
South Dakota RCWP, 132
Spatial information, monitoring, 173–174
Spatial scale, *see* Geographic area/scale
Special Area Management Plan (SAMP), 53
Spray and pray communications strategy, 215
Spring flow, 190
Staffing
   model selection, 191
   planning, 94
Stakeholders, *see also* Partnerships
   beneficial use impairment, defining, 74
   Bennett's hierarchy of evidence, 151
   committee structure, 43
   evaluation process, 147, 148, 150
   failure of previous plans, 53–54, 92, 93, 94
   implementation phase, 17, 102
   institutional arrangements/organization, 40
   management model, 6, 7
   management plan document contents, 125
   monitoring goals, 164
   partnership approach, 10–11
      organizational structure, 45
      participation, 39–40
   planning phase, 17, 120
   pollution control costs, 101
   public participation, 205, 209, 217

   recommend process phases, 11
   restoration issues, 116
   RESULTREACH, 157
   Western Governors' Association principles, 13
   wetland and riparian zone conservation, 114–115
St. Albans Bay, Vermont, 34, 167, 199
Standards/criteria, *see* Criteria/standards; Water quality standards
State agencies/state government, 10
   monitoring responsibility, 174
   partnerships, 40
   technical advisory committee composition, 46–47
Staten Island Blue Belt project, 109
Status report, watershed, 219
Steering committee, 43
   evaluation
      levels of, 153–154
      modification/revision of goals and objectives, 155–156
      targeting information for audience, 160
   leadership of, 43–44
   monitoring reports, 179–180
   rapid resource appraisal (RRA), 45–46
Stewardship approaches, agriculture, 103
Storage tanks, underground, 30
Storming, partnership operations, 50–51
Storm sewers
   data collection, 70
   pollutant sources, 78
   sources of pollution, 2
Stormwater management
   assessment phase, 71
   management plan document funding, 126
   planning process, 100, 108
Stormwater runoff
   build-out analysis, 110, 111, 112
   modeling, 190
   and stream flow, 29
   urban area planning, 108
Strahler method, 25, 26
Strategic plan, lack of, 9
Strategic planning, *see* Planning
Stratified random sampling, 168
Stream banks
   ecological integrity of streams, 72
   erosion
      land use changes and, 79–80
      Lower Big Rib River resource conditions, 88
   mapping, 67
   monitoring, 174–175
   restoration, 116

sediment and, 33
three-tier analysis, 75
habitat structure and function, 22
morphology, flow regimes, 79
restoration, urban area planning, 108–109
stabilization of, 137
sediment sources, 31
Stream channel
flow regime and, 22
geomorphologic processes, 30
management aspects, 31–32
restoration and, 116–117
sediment effects, 3, 33
stages of evolution, 79
urban area planning, 108
urban runoff pollution and, 81
Stream corridors, build-out analysis, 110
Stream flow
assessment phase, 70–71
determinants of, 22
factors affecting quality and function of resources, 21, 22
hydrologic cycle, 29
model selection, 191, 192
runoff and, 79
topography and, 29
Stream network
analysis of individual parts, 73
integration of factors for planning, 86
map information, 66
system overview and linkages, 64
Stream order, 26
defined, 24–25
and watershed size, 66
Stream permanence, and watershed size, 66
Stream quality, impervious cover and, 81, 109
Stressors, three-tier analysis, 74, 75, 76
Structural management practices, 28, 135, 137
Structure, partnerships, 4249
Subregion, 23
Substrate
and habitat structure and function, 22
urban area planning, 108
urban runoff and, 79
Subwatershed, 23
Success, *see* Evaluation
Sulphur compounds, atmospheric deposition, 34
Sunlight, 72
Superior (Wisconsin) Special Area Management Plan, 53
Surface drainage basin, 24
Surface runoff
assessment data, 71
land use and, 32
processes, 28

water cycle, 26, 27
Surface storage
topography and, 29
water cycle, 27
Surface water
and groundwater, 24
partnership requirements, 10–11
regulation approaches, problems created by, 2
Surrogate indices, 86
Surveys
data collection, 69
evaluation process, 144
RESULTREACH steps, 157
techniques, 215–216
Suspended sediment concentrations (SSC), 76
Sustainability, Western Governors' Association principles, 13
SWAT, 186, 191
Swimming, 174
SWMM, 188, 191
Synthesis of information, assessment phase, 86–87, 88

T

Tampa Bay, 34
Target audience, evaluation data, 153, 154
Target identification, modeling data, 187
Targeting programs, 103–104
Targets, planning, *see* Goals/objectives/milestones/targets
Taxation, 115
Teacher training, 219
Team building, 229
Technical advisory committee, 45, 46–47
assessment of watershed, 62
data collection/inventory, 69
Technical assistance
implementation teams, 129–130
plan implementation approaches, 102
Technical limitations, planning constraints, 96
Technical team, channel segment surveys, 79
Technology, Phillips' hierarchy, 100–101
Tell and sell communications strategy, 215
Temperature
assessment data, 68
ecological integrity of streams, 72
urban runoff and, 81
and water quality, 21–22
Template, management plan, 126, 127
Temporal trends, *see* Timeframe
Ten Towns Great Swamp Watershed Management Committee, 44
TetraTech, 184

Index

Thematic mapping, 67
Third-order streams, 25, 26
Threatened species, data collection, 70
Three Rivers Project, Ireland, 207
Three-tier analysis, assessment phase, 74–77
Tillage systems, runoff effects, 136
Tillamook Bay, 177
Time distribution, precipitation, 27
Timeframe, 228
   assessment phase, 58, 59, 63, 68
   evaluation process, 146, 148
   implementation schedule, 130, 131
   local government participation, 58, 59
   management plan document contents, 124
   management practice parameters, 138
   master schedule, 131
   models
      GIS, 199
      selection criteria, 191, 192
   monitoring, 168, 169, 170, 173–174, 178
   planning process, 117–118
   public participation, 224
   reasons for failure of previous plans, 92
Top-down processes, planning phase, 15
Topography
   data collection, 70
   defining watershed, 64
   and flow velocity, 29
   GIS inventory, 199
   map information, 64, 67
Total body contact, designated use definition, 74
Total maximum daily load (TMDL), 78, 100–101
   analysis, 189
   BASINS, 185
   monitoring and assessment issues, 165
   nutrients, 187–188
Total sediment loss, 86
Total suspended solids (TSS), 31, 76, 184
Toxic substances, and water quality, 21–22
Trace metals, urban sources, 33
Tracking, monitoring, 175
Transboundary coordination, 44
Transition zone, landscape approach, 104, 106
Transpiration, 26, 27
Transport
   geomorphologic processes, 30
   infiltration, 30
Transport-related impervious cover, 83
Trend monitoring, 168, 169
Tribal agencies, 10, 44, 227
Tributary plans, 63
Trickle-down approach, management zone, 104
Trust funds
   evaluation process, 143–144
   wetland and riparian zone conservation, 115

Turbidity
   Illinois RCWP project, 33
   three-tier analysis, 74, 76
Turf battles, 53

U

Underground storage tanks, 30
Underscore and explore communications strategy, 215
Unit codes, hydrologic, 23–24
United States
   estimated number and length of river channels, 25
   watersheds by scale/classification, 23
   water resources, 1
United States government agencies
   Army Corps of Engineers water resource projects, 37
   Department of Agriculture Small Watershed Program, 37
   Department of the Interior-Geological Survey, 190
      models, 186
      watershed unit codes, 23–24
   Environmental Protection Agency, 10, 29
      Guidance for Specifying Management Measures for Nonpoint Sources in Coastal Waters, 138
      modeling pollutant loading, 185
      partnership failure, reasons for, 53
      Rapid Bioassessment for Use in Streams and Rivers, 72–73
      Region 5 Water Division Project partnership failure, reasons for, 52
      water quality status assessment, 2–3
      watershed planning efforts, 10
United States Water Resource Council watershed unit codes, 23–24
Units, classes of watersheds, 23–24
Universities, technical advisory committee composition, 46–47
Upland zone, landscape approach, 104, 106
Upper Illinois River, 23
Urban areas/land use, 32, 33
   assessment data, 80–83, 88
   channel changes, 79
   critical areas, 84
   implementation, 134, 135–136
   modeling, 189, 190, 191
   pesticides, 32
   planning process, 107–113
      build-out analysis, 109–110, 112–113, 187

riparian and wetland management, 113–115
pollution sources, 2, 3, 30, 77–78
runoff, 2, 30
    effects of, 79
    Lower Big Rib River resource conditions, 88
    management practice effectiveness data, 138–139
    as resource, 100
    volume of, 31
Usability of program, evaluation data, 152
USLE, 192
USLE/MUSLE, 186
Utility, valuation standards, 158

V

Validation, model, 189, 190
Values, 227
    conflict resolution, 55
    failure, reasons for, 54
    local government participation, 58
    social capacity building, 218
Variable source area process, 28
Vector-based GIS, 194–195, 196
Vegetation
    and habitat quality, 79
    and habitat structure and function, 22
    mapping, 67
    restoration, 116
    urban area planning, 108
Vegetative practices, 137
Verification, model, 189, 190
Vermillion River, Illinois, 34
Vermont RCWP Project, 34, 99, 167, 199
Violations, permit, 146
Visual inventories, 46, 68
Volunteer contributions
    implementation, 139
    monitoring, 169, 176–177

W

WASP, 190
Wastewater, 30; *see also* Runoff
    build-out analysis, 110, 112
    injection into groundwater, 30
Waste water treatment facilities
    build-out analysis, 113
    pollutant sources, 77–78
    urban area plan requirements, 107
Waterbody
    analysis of individual parts, 73

integration of factors for planning, 86
system overview and linkages, 64
Water budget, assessment data, 71–72
Water cycle, *see* Hydrologic cycle
Water discharge, data for assessment, 68
Water quality
    analysis of individual parts, 73
    assessment components, 61, 70, 73
    ecological integrity of streams, 72
    factors affecting, 21–22
    goals, management plan document contents, 127
    integration of factors for planning, 86
    Lower Big Rib River resource conditions, 88
    management plan document contents, 123, 126, 127
    model calibration/validation, 190
    monitoring, 173
        sources of variability, 171
        trends in, 168
    system overview and linkages, 64
    urban runoff and, 79, 81
    wetland-to-watershed ratio and, 113
Water quality data
    assessment reports, 70
    model calibration/validation, 190
Water quality standards
    assessment phase, 61
    management plan document contents, 123
    monitoring, 173
Water quantity, assessment components, 73
Water regime, factors affecting quality and function of resources, 21, 22
Water resources, data collection, 70
Watershed monitoring, 164–167; *see also* Monitoring
Watershed processes, 21–35
    aspects for management, 31–32
    assessment phase
        definition of, 65–67
        scale of, 63
        strategy development, 100–101
        synthesis of interacting factors, 86
    atmospheric deposition, 34–35
    characteristics of, management aspects, 31
    defined, 3
    drainage network, 24–25, 26
    geomorphologic processes, 30–31
    land use, 32–35
        agriculture, 33–34
        urban, 33
    map information, 66
    units, classes of watersheds, 23–24
    water (hydrologic) cycle, 25–30
        groundwater, 29–30

Index

precipitation, 26–28
runoff, 28–29
stream flow, 29
water quality, factors affecting, 21–22
Watershed scale modeling, 189
Watershed scale monitoring, 175–176
Watershed treatment model (WTM), 186
Watershed units, scale/classification, 23–24
Water supplies
data collection, 70
designated use definition, 74
urban area planning objectives, 107
Water table drawdown, 30
Water usage, agricultural, 83
Waterville, Ohio, 147–148
Waukegan River Project, 188
Weather, 66, 68, 174
Wellhead area, defining watershed, 64–65
Well water, monitoring, 174
West Branch Delaware River Model Implementation Project, 166
Western Governors' Association (WGA) Principles for Environmental Management in the West Policy Statement, 12

West Fork of Madison River, Montana, 79
Wetlands
build-out analysis, 110
loss of in urban areas, 33
planning process, 113–115
watershed-to-wetland ratio, 22
Whatcom County, Washington, 133
White Clay Lake, Wisconsin, 71, 166
Willow Creek, 137
Wisconsin Nonpoint Source Program, 171
Work plan, 46
Workshop, kickoff, 39
World Commission on Environment and Development, 11
Worst-first approach, 104

Y

Ythan River, Scotland, 99–100

Z

Zoning, urban area planning, 108